THE TELEVISION AUDIENCE: PATTERNS OF VIEWING

The Television Audience:

Patterns of viewing

G.J. GOODHARDT
Thames Polytechnic
A.S.C. EHRENBERG
London Business School
M.A. COLLINS
*Social and Community
Planning Research*

and

Aske Research Ltd

SAXON HOUSE

Published by

SAXON HOUSE, Teakfield Limited,
Westmead, Farnborough, Hants., England

Reprinted 1978, 1979

ISBN 0 347 01102 0

Printed and bound in Great Britain by
Ilfadrove Limited, Barry, Glamorgan, S.Wales

Contents

Foreword   vii

Preface   ix

Detailed Contents   xiii

1   TELEVISION PROGRAMMING   1

2   AUDIENCE FLOW   9

3   CHANNEL LOYALTY   23

4   OTHER FACTORS   37

5   REPEAT-VIEWING   55

6   INTENSITY OF VIEWING   75

7   AUDIENCE FLOW IN THE US   85

8   AUDIENCE APPRECIATION   97

9   THE LIKING OF PROGRAMME TYPES   109

10   TELEVISION AS A MEDIUM   125

Appendix A: Audience Measurement   136

Appendix B: Updating to 1974   144

IBA Reports   148

Further Readings   150

References   151

Index   155

Foreword

Where facts are few, myths abound. There can be few better areas for illustrating this observation than television. Despite the collection of enormous masses of data on a continuous basis over many years by both broadcasting organisations in the United Kingdom, systematic studies of how the viewer actually *behaves* — the pattern of his viewing — have hitherto been remarkably scarce. Professor Ehrenberg and his colleagues have taken a first but important step towards rectifying this state of affairs.

This book has arisen out of a programme of research undertaken for the Independent Broadcasting Authority by Goodhardt, Ehrenberg and Collins of Aske Research Ltd, London, which started in 1967 and is still continuing. The basic aim throughout has been to get beyond the detailed and specific information supplied by audience measurement studies, such as those by Television Audience Measurement Ltd (TAM) and Audits of Great Britain Ltd (AGB) for JICTAR, to more generalised findings about viewer behaviour. The fact that the data are derived from the same panel of viewers each day makes it possible to do so, and in the process various assumptions that have been apt to be taken for granted are subjected to critical scrutiny.

Are there identifiable groups of viewers who express common programme preferences, and who actually exhibit similar viewing behaviour patterns? Are viewers demonstrably "loyal" to a channel, and if so does this vary from ITV to BBC1 to BBC2? Can the audience be "caught early" by clever arrangement of the schedule, and then "held throughout the evening"? What proportion of the audience to a programme on a given channel will see a different programme on another day? Will it be greater if the second programme is on the same day at the same time a week later, or if it is linked in some way, as a complementary programme, as part of a series, or as another episode in a serial?

All who have been concerned with television have heard confident and seemingly authoritative assertions about these and related questions which are typical of the subject-matter covered in *The Television Audience*. These are important questions because they are the very stuff from which programme, scheduling and broadcasting policy decisions are created. It is probable that the implications of some of the findings in the book will come as a surprise to its readers: it is hoped that they will be helpful, in the future, when such matters are discussed, as they certainly will be.

The Television Audience contains material which will be of interest not only to broadcasters but also to those who, while outside the area of active broadcasting, are concerned with the media in general and the social issues associated with mass communication. Research workers will also find it instructive as a demonstration of making fuller use of data, extensively and expensively acquired and then not always studied as deeply as they should be.

The authors have made a significant contribution to knowledge in an important area, and at a time when the issues which are raised are matters of public concern and argument. I commend this book to the thoughtful consideration of the reader.

Ian R. Haldane
Head of Research
Independent Broadcasting Authority

Preface

The average family in the United Kingdom currently watches television for more than five hours a day. *Individuals* on average watch almost three hours or so a day. In the great majority of homes the television set is therefore switched on for most of the evening. A similar pattern occurs in the United States and much of Western Europe.

With television occupying such a significant part of the leisure time of many people, it is not surprising that the medium has become a subject of major social and political concern and even controversy. Much of the discussion has centred on the likely *effects* of television. How does a heavy diet of violent programmes affect children? Does the heavy viewing among working class households add to their cultural deprivation? Could television be an effective educational vehicle? More operational questions concern the number of different TV channels, how they should be run, and so on. Various studies on both sides of the Atlantic and elsewhere have tried to deal with these and similar problems.

Many of these studies have been somewhat disappointing, being piece-meal in their analysis and unconvincing in the generality of their conclusions. One possible reason has been that the studies usually failed to deal with perhaps the most basic question of all: the sort of television viewing that people actually do. Before we can learn about the effects of particular types of television on people, we need to take into account what they watch.

Implicit in certain criticisms of programmes on crime and violence is for example the thought that such programmes tend to attract the same type of regular viewer or "addict", on whom they then have a harmful effect. As will be shown in this book, it is easy to explore the first premise – that programmes of a given type attract a particular group of people. Until this is determined, the effects of such programmes can hardly be realistically studied.

The main purpose of this book is in fact to describe how people view. We examine their loyalty to particular programmes or types of programmes, their loyalty to particular television channels, and the nature of their switching between channels and programmes.

We also examine data on the audience's *appreciation* of programmes. Most people agree that it is not enough to assess television programmes solely in terms of the size of the audiences they attract (the *ratings*). Aside from questions of quality versus quantity, audience size is an inadequate measure of a programme's inherent appeal because ratings are affected by the programmes on alternative channels, by the views of different family members, by time of day, and so on.

Most of the findings in this book are simple. Patterns are found which tend to be regular and hence become predictable. Such regularities can therefore become a basis for understanding the medium, for making forecasts about the nature of the audiences for different programmes or programme schedules, and for testing ideas about the impact of different forms of television policy. Since our own expertise lies more in the study of viewing patterns than in the wider issues of policy, the book itself concentrates on the audience's viewing behaviour as such.

Television is a subject on which many people have views and preconceptions. On perusing an earlier account of one of the findings here, one lay-reader concluded that the report seemed to be saying:

> A person who has just been watching a detective-type programme on one channel will tend to switch to another channel if that is then showing another programme of the same type.

On being told the report actually showed the *opposite* effect (i.e. that there was *no* such special tendency), the reader replied:

> That is what I actually *thought* it said, but I couldn't believe it.

Correcting widely held misconceptions is one practical value of the results of systematic research, although this often takes time and effort.

The structure of the book

Television viewing is naturally influenced by the programmes offered. In Chapter 1 we therefore outline the nature of television programming and audience measurement procedures. (The latter are described more fully in Appendix A.)

Chapter 2 introduces the main concepts of "audience flow" – the extent to which different programmes, or different episodes of the *same* programme, have viewers in common. The basic "duplication of viewing law" is introduced here. Chapters 3 and 4 develop more detailed aspects of audience flow: programme choice, channel loyalty, the overlap between the audiences of different programmes, and the factors involved.

Chapter 5 examines repeat-viewing, i.e. how many viewers watch more than one episode of the same programme, and how this builds up for a longer series of episodes. Chapter 6 broadens this to an analysis of the varying amounts of television which different people view. The incidence of heavy and light viewers serves to explain the nature of audience duplication between different programmes.

Chapter 7 extends the results in the UK to viewing behaviour in the US. Although the range of data covered is less wide and some of the conditions different, the main results still hold.

Chapters 8 and 9 move from viewing behaviour to viewers' expressed appreciation or liking of the programmes seen. Audience appreciation is examined in relation to the size of the audience and the percentage of repeat-viewers attracted by different programmes. The fact that some people like different programmes of the same type is reconciled with the apparent absence of such preference groupings in people's actual *viewing* behaviour.

Chapter 10 draws together the various findings and briefly sets out some implications for the nature of television as a medium, although this is a monograph on our analyses of the television audience rather than an attempt to explore the medium in general terms.

The bulk of the results described was built up from our work in the UK, covering data mostly from the late sixties to 1971. To provide updating and confirmation, a check of the findings on viewer behaviour has been made using data for 1974; the results have been added as Appendix B.

Acknowledgements

Most of the findings described in this book arose from studies commissioned by the Independent Broadcasting Authority from Aske Research Ltd (as listed at the end of the book). We are grateful to the IBA and to the American Research Bureau (Arbitron) for permission to quote from reports written for them, and to JICTAR, AGB Ltd, TAM Ltd and Leo Burnett Ltd for permission to quote from their data.

We also wish to express our appreciation to the IBA for supporting the preparation of the book itself, to Peter Doyle, Professor of Marketing at the University of Bradford, for his work on the initial draft, to Mrs Myra Davies of Typlan for typing the tables, and Mrs Helen Bloom Lewis for her many helpful comments. We are however particularly indebted to Dr Ian Haldane, Head of Research at the IBA, for his encouragement and support throughout the whole programme of research.

G.J.G., A.S.C.E., M.A.C.
London, June 1975

Detailed Contents

1   TELEVISION PROGRAMMING   1

Historical Background   1
The Programmes Available   4
Audience Measurement   6
Summary   8

2   AUDIENCE FLOW   9

Characterising the Audience   9
Viewing Patterns   10
Audience Duplication   10
Repeat-Viewing   15
Rating Levels   17
Some Statistical Considerations   17
Summary   19
Mathematical Appendix   20

3   CHANNEL LOYALTY   23

The Main Steps   23
Two ITV Programmes   24
Other ITV Programmes   25
The Duplication of Viewing Law   26
Viewing on Non-Consecutive Days   28
Channel Switching   29
High Duplication on BBC   31
Summary   35

4   OTHER FACTORS   37

Demographic Factors   37
Changes Over Time   38
Viewing at Off-Peak Times   39
Week-End Viewing   42
The Inheritance Effect   43
Programme Types   46

A Specific Application: News Broadcasts 49
Summary 54

5 REPEAT-VIEWING 55

"Z-Cars" 55
55% Repeat-Viewing 56
Irregular Programming 61
Viewing the Other Channel 62
Non-Consecutive Episodes 63
"Mr Trimble" 64
Audience Cumulation 64
Summary 71
Mathematical Appendix 71

6 INTENSITY OF VIEWING 75

Hours Viewed 75
Channel Choice 76
Programme Choice 77
Repeat-Viewing 78
Duplication of Viewing 79
Summary 82

7 AUDIENCE FLOW IN THE US 85

Repeat-Viewing 86
Duplication between Different Programmes 89
The Inheritance Effect 93
Summary 95

8 AUDIENCE APPRECIATION 97

The Appreciation Index 98
Appreciation Index and Audience Size 99
Appreciation Index and Repeat-Viewing 100
Appreciation Index and Individual Repeat-Viewing 101
"Coronation Street": A Case History 104
Programmes of the Same Type 105
Summary 106

9 THE LIKING OF PROGRAMME TYPES 109

Programme Clusters 109

Programme Character   115
The Nature of Programme Clusters   117
Reconciliation with Viewing Behaviour   119
Summary   122

10   TELEVISION AS A MEDIUM   125

Simple Findings   125
Some Broad Implications   127
The Effects of Television ――――   131
Summary   135

Appendix A: AUDIENCE MEASUREMENT   136

Measurement Procedures   136
Presence and Attention   140
Audience Reactions   142
Summary   143

Appendix B: UPDATING TO 1974   144

Repeat-Viewing   144
Audience Duplication   145
Inheritance Effects   146
Other Work   147
Summary   147

IBA Reports   148

Further Readings   150

References   151

Index   155

1 Television Programming

Every year millions in money and man-hours are poured into television —
its technology, programmes and advertisements. The object of all this
effort is to reach and communicate with the audience — to entertain it,
stimulate it, inform it, educate it, influence it in some way.

Yet as integral a part of modern life as television has become, there is still
no satisfactory way to evaluate the success or failure of these efforts. What
use do viewers make of the channel and programme choices open to them?
How much satisfaction do they derive from television? What needs or wants,
however defined, remain unfulfilled? Do advertisements really sell? Do they
make us buy things we do not want? How great is television's influence on
the audience and on society in general? Is it advantageous or detrimental?
Can a preponderance of violent programmes really warp some minds?

The answers to these and other questions matter not only to those
working in television and advertising, but also to the authorities
responsible for the medium and to the public as viewers and consumers.
But the questions can hardly begin to be answered until one knows how
the audience in fact behaves — the patterns of people's viewing.

Hence the purpose of this book — to build a general picture of the
regularities that exist in the behaviour of viewers and in their attitudes
towards programmes, and in so doing to contribute to a better understanding of the medium and to demolish some of the shibboleths that
have found their way into popular belief. Our topic here is relatively
narrow — the television audience as such, rather than the broader
interpretative issues of the role of television in society. Most of the studies
that form the basis of this book were carried out in the UK. We therefore
concentrate here on British viewing patterns. But the general approach
and many of the fundamental results will apply also in other countries, as
is illustrated for the US in Chapters 2 and 7.

Historical background

The United Kingdom

Television transmissions began on a regular basis in the UK in November
1936. By the outbreak of World War II there were about 25,000 sets in
use; now there is virtual saturation with little short of 20 million sets.

1

The UK had only one channel until 1955. This monopoly was held by the BBC (the British Broadcasting Corporation), a public corporation financed from revenues derived by issuing annual licences for sets to the public.

In 1954 the government established another public corporation, the Independent Television Authority, now called the Independent Broadcasting Authority or IBA. This currently appoints fifteen regional independent television (ITV) programme companies in fourteen areas to provide broadcasts on a second channel. ITV is financed entirely by advertising revenue.

In 1964 parliament awarded another channel to the BBC. The intention was that BBC1 should aim at majority interest groups and BBC2 at various special interests. Since January 1968, colour transmissions have been developed and used on all three channels.

Subject to physical problems of adequate local reception, the typical viewer in the UK therefore has three channels to choose from — BBC1 and BBC2, each of which generally broadcasts its programmes nationally, and ITV, which generally broadcasts different programmes in each of its regions, although some are "networked", i.e. shown simultaneously in two or more regions. (People in "overlap areas", at the edges of two ITV regions, mostly have the choice of tuning in to either ITV region, often giving them a choice of four programmes.) ITV and the BBC tend to split the total TV audience about 50:50. Of the two BBC channels, BBC2 tends to attract fewer viewers.

Programming policy by BBC1 and BBC2 is "complementary", aiming to provide a varied choice at any one time. Thus they generally stop and start programmes together, to facilitate viewer switching between channels. ITV programming on the other hand is largely competitive, aiming to maximise its share of the audience vis-à-vis both BBC channels. ITV tends to put on a similar rather than a different programme to that on BBC1 at a given time and to schedule programme timings differently from the BBC to reduce the likelihood of switching ("non-coterminous programming").

The amount of advertising material that can be shown on ITV in the UK is controlled by law, with not more than seven minutes allowed in any one clock-hour and a maximum daily average of 6 minutes per hour. In other countries the situation differs. In Germany, for example, advertising is concentrated into a half-hour time-band from 7 to 7.30, and France uses a similar system.

The United States

We now briefly describe television in the US, to provide a contrast to the UK and also a background to the US findings in Chapters 2 and 7.

The effective start of TV transmissions in the US was delayed by commercial competition (as contrasted to the initial government monopoly in the UK) involving differences in broadcasting techniques and types of receivers used. In 1941 the Federal Communications Commission (FCC) — a federal agency that awards broadcasting licences and oversees the broadcasting media — authorised full commercial operation on 18 VHF channels. However, debate continued on three different synchronising methods. Television broadcasting was limited to only six experimental stations during World War II and licensing did not resume until 1945. Further delay followed while the FCC and the infant TV industry debated the merits of using monochrome and VHF channels versus colour and UHF. An FCC verdict decided in favour of the former in 1947, allocating 12 VHF channels to television and the American TV gold rush was on.

By 1948 there were 41 stations serving 23 cities, regular network services had started serving the Midwestern states as well as the East Coast, and important advertisers had begun experimenting with the medium and sponsoring large scale programming. The East and West Coasts were joined by a coaxial cable and microwave network in 1951, allowing national TV network services that soon reached 60% of all American homes. There was no doubt of the medium's popularity — by 1950, 10 million sets had been sold and by 1968 this number had risen to 90 million. The number of homes with at least one set is almost 70 million. An increasing number of homes have more than one set.

There are over 200 separate "markets" covered by their own separate TV stations. In some of the largest, like New York and Los Angeles, people can receive six or eight or more stations (depending on their location). In 1973 there were in all more than 900 TV stations on the air, of which about 230 were non-commercial, made up mainly of a national education network and various community stations (all of which tend to attract small audiences). Non-commercial stations receive sporadic funds from the federal government but are basically financed by grants from outside organisations, public service advertisements and viewer contributions.

The 700 commercial stations are financed by advertising revenue. In most markets three stations are affiliated to the three national network companies whose programmes are generally broadcast simultaneously across the country (but bearing in mind that different time-zones exist). These three networks, CBS (Columbia Broadcasting System), NBC (National Broadcasting Company) and ABC (American Broadcasting Company), have played a key role in the development of the US television

3

industry. They tend to dominate the ratings and are highly competitive towards each other. They are less distinguishable for their programme content than are the BBC and ITV in the UK. In addition there are 20 regional networks and various independent commercial stations.

There is no law or FCC ruling governing the amount of advertising that may be carried on US television stations, but the National Association of Broadcasters has a voluntary code that specifies a maximum of 16 minutes per hour.

The programmes available

On each British channel a variety of programmes tends to be screened — comedies, westerns, crime stories, plays, old films, sport, news, some documentaries or other informative programmes, and so on. The selection and mixture will vary with the aims and financial viability of the programme company or station. A major question is the balance of programmes, e.g. what is the "best" number of westerns and situation comedies or of crime programmes? Where advertising provides the revenue, a broad tendency is to aim at maximum audiences, subject to some public accountability or control for balance by the IBA (and the need to obtain a renewal of one's licence to broadcast). But even a broadcaster like the BBC which is independent of advertising revenue is conscious that one important measure of its success is the size of the audiences it can attract and hold.

Programme types

A broad comparison by type of programme brings out certain differences in the "programme mix" of public service and commercial channels. Table 1.1 gives a comparison for one public service and one commercial channel in the US, and the three channels in the UK (adapted from Williams, 1974).

The definitions of the programme categories are usually self-evident and here we are only interested in the broad patterns. "Series" in Table 1.1 refers to drama programmes (westerns, crime, situation comedies, etc.) where certain regular characters appear in successive episodes with self-contained plots, while "Serials" additionally have a continuing story-line. "Education" refers largely to programmes for schools and colleges, plus a small proportion of adult education shows. Young children's educational broadcasts are listed under children's programmes.

Table 1.1

The distribution of programmes on various channels
(by time of duration)

March 3-9, 1973	Commercial		Public Service		
	ITV UK*	Ch. 7 US**	BBC1 UK	BBC2 UK	KQED US**
Programme Type	%	%	%	%	%
Series and Serials	17	17	7	4	5
News & Public Affairs	13	14	25	12	22
Movies	12	18	6	11	6
Education	12	2	23	29	26
Commercials	11	14	-	-	-
Gen. Entertainment	10	24	7	7	0
Childrens' programmes	8	4	11	6	27
Features and Document.	6	1	7	20	6
Sport	6	4	6	2	2
Plays	3	0	5	5	0
Publicity (internal)	1	1	1	1	1
Religion	1	1	1	0	0
Arts and Music	0	0	1	3	5
Total hours transmitted	103	133	100	62	94

* Anglia ** San Francisco

"General entertainment" includes musical shows, variety shows, games and quiz shows, and those talk shows which in manner and matter are more linked to "show-business" than to public affairs discussions (for further details see Williams, 1974).

Repetitive programming

Perhaps the dominant feature of television programming is that it is repetitive. The same programmes – or strictly speaking, different *episodes* of the same programme – tend to be shown at the same time each week, with perhaps three or four major upheavals in the programme schedule a year.

Some of this repetitive programming takes the form of *serials* with on-going story-lines, and a strong element of continuity. Others are *series*, film or drama slots, regular current affairs programmes, week-end sports broadcasts, and of course the news, where the need to watch regularly would appear to be less.

Repetitive programming superimposes a firm structure on television. The number of occasions where there is something radically new to watch are relatively few. There are several reasons for this.

Firstly, television demands an enormous amount of programme material. Probably the only way to cope with it at a reasonable level of technical competence is to develop formula shows and stereotypes. For everything to be original and new would require the development and organisation of experienced talent that is hardly likely to be available. Proverbially the film "Lassie" was followed by several television series, then by "Son of Lassie", and so on.

Secondly, we *like* repetition. As viewers, we find it convenient to know in advance when programmes are being shown — even if one is only going to miss them ("Tonight is the second time I can miss seeing so and so".) But our liking for repetitive broadcasting goes well beyond the convenience of regular and predictable timing to the fact that we learn to appreciate certain characters, comedy routines or stereotype plot situations through familiarity. We develop habits and preferences.

A third reason for repetitive programming on *commercial* television is that fairly stable and predictable audience levels are required at each point of time so that advertisers have some idea and even guarantee of what future audience sizes they are buying. Typically, programming in the UK on ITV tends to be somewhat more rigid than that on the BBC.

Repetitive programming also occurs at closer than weekly intervals. The outstanding example is, of course, the daily news. But some series in the UK are seen twice weekly or more often, especially some on week-day afternoons for young children in specific age groups. In the US, the smaller "independent" stations, and also sometimes the networks during day-time, go in for "strip-programming". This means that episodes of the same programme (often old ones) are shown regularly on each of the week-days (e.g. "The Lucy Show", "The Flintstones" and "Star-Trek", five times a week each).

Despite this largely repetitive nature of TV programming, little seems to have been published about the extent to which viewers of one episode of a programme also view the next, or what factors determine high or low repeat-viewing loyalties. Yet as a measure of audience response or appreciation of programmes the level of repeat-viewing might seem potentially more valid than the mere size of the audience. This is discussed at various stages of this book, but especially in Chapter 5.

Audience measurement

The analyses of the television audience discussed in this book are generally based on routine audience measurement data. Broadcasting is unique in the extent to which it has to rely on market or opinion research to indicate its reach. Other media and industries may use circulation figures, box-office receipts or ex-factory shipments to estimate sales or the number of customers. But for television the only currently practical way to determine which programmes are heavily viewed or favourably evaluated is to conduct sample surveys among members of the audience. This produces the ratings, i.e. measures of the percentage of the total population viewing.

The main measurement procedures are briefly outlined below. A fuller discussion is given in Appendix A.

Audience size

There are two separate audience measurement systems in the UK, each covering viewing on all channels but differing radically in technique.

The BBC interviews some 2000 people a day, but a different sample each time (adding up to well over half a million people a year). Each informant is asked only about one day's viewing; hence such data provide no information about the flow of the audience between programmes on different days or in different weeks.

The main audience measurements by the independent television interests are sponsored by JICTAR (the Joint Industry Committee for Television Advertising Research, made up of ITV programme companies, advertising agencies and advertisers). The actual work is currently carried out for JICTAR by AGB (Audits of Great Britain).

These data are based on panels of households whose viewing is measured more or less continuously over successive days and weeks. Sample sizes tend to be from 100 up to 350 homes per region, totalling about 2600 nationally and comprising nearly 8000 people. Individuals' viewing is measured in quarter-hour time-bands by weekly diaries, the qualifying definition being that one has viewed for at least 8 minutes in the quarter-hour. The diary data is monitored by an electronic meter attached to the TV set, which itself give the minute by minute "set-on" ratings which are the most widely quoted audience measures. These procedures tend to give reliable results. The quarter-hour measurements of individuals' viewing are the data mainly used in this book.

In the US the audiences of the national network programmes are measured by a meter panel operated by the A.C. Nielsen Company and supplemented by separate diary measurements of individual viewers' behaviour. Audience ratings for local "markets" are measured by both the American Research Bureau (Arbitron) and Nielsen using one-week diaries.

Audience appreciation

Audience size by itself is an incomplete index of viewers' reaction to the programmes on offer. In Chapters 8 and 9 we discuss results from more attitudinal measurements, e.g. from "Audience Appreciation" panels operated by the IBA. Panel members report in alternate weeks with mail diaries, assessing the programmes they have viewed in terms of an overall Appreciation Index (running from "Not at all interesting and/or enjoyable" to "Extremely interesting and/or enjoyable"). Special scales for selected programmes and individual comments are also used.

The BBC also employs regular monitoring of audience reactions but uses different techniques. These and more *ad hoc* research into television audiences are outlined further in Appendix A.

Summary

In developed countries like the UK and the US, most people can choose from different television programmes shown on two or more channels, each tending to carry a variety of different types of programme. When the channels compete for audiences, they may screen similar programmes at the same time, thus reducing the effective choice open to the viewer. *Complementary* programming offers different types of programme at any given time. This is used by BBC1 and BBC2 for example.

The dominant feature of television programming is probably its repetitive nature, with different episodes of the same programme usually being shown at the same time in successive weeks.

2   Audience Flow

"Audience flow" is a term that describes two main concepts. One is the extent to which viewers of a particular programme are also viewers of another programme ("audience duplication"). The second is the extent to which viewers also watch another episode of the same programme, usually shown a week later ("repeat-viewing"). The study of such viewing patterns is fundamental to our understanding of the way people use the programmes they are offered.

Characterising the audience

In this book we characterise the audience of a particular TV programme by what *other* TV programmes they also watch. The reasons for this approach are several.

First and foremost, it seems relevant and meaningful to ask whether the people who watch a certain western also watch many other westerns, but *not* current affairs programmes say. That tells us a lot about the kind of viewers they are. Secondly, data for such analyses exist in great profusion so that any findings can be well substantiated. Thirdly, a range of simple and generalisable results about viewing patterns have in fact been established. Finally, the same approach and many of the conclusions obtained are readily applicable in the future and in other countries.

However, there are also many other possible ways of describing the audience to a particular television programme — for example by the viewer's age, sex, social class, occupation, size of household, number and age of children, ownership of a motor car or cheque book, and so on. Or again, one may ask the viewers attitudinal questions, e.g. about the state of the world or about television. Viewers can also be differentiated by measures of their personalities, their tendencies towards violence or sloth (both possibly induced or aggravated by television), and by their other activities and interests (e.g. reading of print media and leisure habits).

These other approaches have been followed in a number of studies of television audiences (see the Further Readings listed at the end of the book). For example, it has been found that women in the UK watch somewhat more TV than do men, and that upper and middle class people watch somewhat less than members of the working classes. But these approaches have seldom produced any particularly revealing or insightful results and thus will not be pursued extensively here.

Finally, viewers can be questioned about their appreciation or liking of particular television programmes, as will be considered in Chapters 8 and 9. But such attitudinal responses become much more telling if their interpretation is linked to what people actually *do*, by way of choosing programmes to watch.

Viewing patterns

Viewing patterns may be thought difficult to interpret because so many different factors can influence them. For example, if only 52% of the viewers of a certain western also watch the Friday "News at Ten", this could be because some of them were not at home then or had gone to bed, because of alternative programmes on other channels, because of what other members of the family or guests were doing, because there happened to be no particularly newsworthy news that day, or even because some of them do not actually "like" watching the news. So how do we judge this figure?

If we were told that 52% of the viewers of the western read *The Times*, we might find that easier to interpret, and indeed surprising. We would already know that such a figure is very high compared with the percentage of the population as a whole who read *The Times*, and also that readers of *The Times* tend to be "serious" readers. In contrast, it generally would not be immediately obvious whether a figure of 52% of the western audience watching the news should be considered high, or low, or perhaps just normal. We do not know what other types of programme the typical audience of the news watches anyway.

Thus what is needed in the first place are some interpretative "norms", i.e. typical patterns of audience duplication and repeat-viewing for different programmes. Once we determine these, individual figures like the 52% should be easier to interpret. It is our concern in the next few chapters to describe and discuss such general patterns.

Audience duplication

A basic aspect of viewing behaviour is the extent to which viewers of one programme also watch another programme. This may be screened on a

10

different day and perhaps on a different channel. The programme may be of the same type or of a different type (e.g. viewers of a crime series watching either another crime programme or a comedy show).

The question is, "To what extent are any two programmes watched by the same people?" There are many factors involved — day of week, time of day, channel, type of programme, programme content, audience size or "rating" of each programme, and so on. What are the patterns and the factors that influence them?

One practical question is whether certain types of programmes, such as westerns, or crime series, or documentaries, attract viewers with special preferences for that type of programme. Such a belief is widely held. To determine whether it is correct, it needs to be established whether the viewers of one crime or police programme are more likely than other people to see *other* crime programmes.

Sufficient research has already taken place to provide relevant results about audience duplication patterns. One basic result is known as the duplication of viewing law.

The duplication of viewing law

The major influence on the level of audience duplication between two programmes is usually the rating level or audience size for each programme. This is to some extent self-evident. If almost nobody watches a programme, audience duplication with any other programme will be low!

But the effect of rating levels is pervasive even in less extreme cases. It is the underlying feature of many of the results discussed in the next chapters and is exemplified by the duplication of viewing law. This states that

> The proportion of the audience of any TV programme who watch another programme on another day of the same week is directly proportional to the rating of the latter programme (i.e. equal to it times a certain constant).

Thus, given knowledge only of the ratings and the value of the constant, we can generally predict what proportion of the audience of one programme will also watch a second programme on another day: the audience overlap varies with the rating level (the audience size) of the second programme.

For example, if the constant or "proportionality factor" for ITV programmes is 1·4 and the rating of the second programme is 20 (i.e. 20%

of the total population watched it), then the duplication law states that about

$$20 \times 1 \cdot 4 = 28\%$$

of the audience of the *first* programme will have watched the second programme. (The proportionality factor here is the same for most types of programmes, i.e. a constant. But it differs according to the channels on which the programmes are shown. More viewers of one programme watch another if the two programmes are on the same channel rather than on different ones, even though we are talking here of programmes shown on different days).

The numerical example brings out the dramatic feature of the result. It shows that the proportion of the audience of *any* ITV programme one day who saw the second programme on the other day is about 28%, i.e. that the audience overlap generally depends only on the proportionality coefficient (here 1·4) and the programme's rating (here 20), and *not* on programme content. This will be shown in detail in the following chapters.

The result might seem surprising, e.g. that the programme is seen by 28% even of the audience of a children's programme shown on some previous afternoon. But if we are talking about adults, few would have watched the children's programme, and about 28% of these might well then watch the other programme in question. It is important that what the duplication law refers to should be made clear. The population that is being analysed may be all adults in homes able to receive TV, or some subgroup like housewives, or younger men, or whatever.

For some purposes it is helpful to express the duplication law more symmetrically, i.e. not as the proportion of the audience of programme A, say, who also watch programme B, but as the percentage of the whole population who watch both A and B. The law then takes the form:

% of population who watch both A and B
= rating of A × rating of B ×
the proportionality coefficient divided by 100

(The mathematics of these different formulations is set out in the appendix to this chapter.)

Thus if programme A had a rating of 30, and programme B one of 20 as before, then with a proportionality factor of 1·4, about

$$\frac{30 \times 20 \times 1 \cdot 4}{100} = 8 \cdot 4\%$$

of the population will have watched *both* programmes. (The divisor of

100 is needed if we are using percentages and not proportions.) To express this common audience as a proportion of the audience of programme A, we divide 8·4 by 30 (A's rating) to find that 28 or 28% of the viewers of A also watched B — the same figure as we had before.

Empirical basis

The duplication of viewing law may be difficult to accept at first sight. But it holds under many different circumstances, i.e. it is an empirically based generalisation. It also has a theoretical basis or explanation which will be discussed in Chapter 6, but the main point remains that the law describes to a close degree of approximation what actually happens. One of the tasks of this book is to illustrate how and when the law works, and also to pinpoint and discuss those situations where audience flow takes *different* forms.

The law generally accounts for the major single factor in audience duplication — audience ratings. The influence of other factors shows itself as deviations from the predicted results, and such deviations tend in most cases to be small. But they could not be isolated without first eliminating the effects of the rating levels as such.

However there are situations where the law does not apply, at least not in its direct form. "Availability" is one factor. Thus, the audience overlap for late-night programmes is consistently higher than the duplication law predicts because the same people tend to stay up late. Again, two programmes shown in succession on the same channel also have a much higher degree of audience duplication than the law predicts because of the "inheritance effect", as discussed in Chapter 4. In its present formulation the law also tends to overstate the duplication for programmes with very high ratings. These rarely occur, but some mathematical reformulation will be required in due course.

Accepting the duplication of viewing law is therefore not a consequence of any purely theoretical argument or assumption, but only something we need to do in as far as it fits the facts.

An illustration: "Arts Documentaries"

We now illustrate how a result like the duplication of viewing law can be used to test whether certain types of programmes attract particular types of habitual viewers. If such a tendency were to exist for a certain programme type, then the actual proportions of the population watching a pair of such programmes should be higher than the predicted value given by the duplication law.

13

For example, do "Arts Documentaries" (using the formal IBA definition of programme type categories) attract a steady following of "cultured" viewers? In the first week of May 1967, five such programmes were offered on the ITV and BBC1 channels, as shown in Table 2.1.

Table 2.1

Viewing of "Arts Documentaries"
(adults, May 1967)

Channel	Time (pm) & Day		Programme	Rating
ITV	9.15	Wednesday	Cinema	43
ITV	10.15	Friday	This Week – the Arts	12
BBC 1	11.00	Wednesday	Masterwork – Piano	1
BBC 1	10.15	Sunday	Contours of Genius	5
BBC 1	11.00	Sunday	Look of the Week	2

The duplication law predicts that the percentage of the total adult population who watched both "Cinema" and "This Week – the Arts" (both ITV programmes) is the product of the audience ratings multiplied by the *within-channel* constant, at that time 1·4, divided by 100, i.e.

$$\frac{43 \times 12 \times 1\cdot4}{100} = 7\cdot2\%$$

(The divisor of 100 is needed again here because all the figures are expressed in percentages.) Examination of actual viewing data (measured then by Television Audience Measurement) showed that in fact 8·0% of the population had watched the two programmes. This is close to the prediction of 7·2, with a discrepancy of only 0·8, or 1 to the nearest whole number.

It may be thought that "Cinema" and "This Week – the Arts" are rather different in terms of content. But similar comparisons of the observed and predicted audience duplication values can be made for all the other pairs of programmes in the broad "Arts" category. For example, the actual duplication for ITV's "Cinema" and the BBC's "Contours of Genius" was about 1 rating point below the estimated BBC × ITV norm. The duplication for ITV's "This Week – the Arts" and "Contours of Genius" was about 1 rating point *above* the predicted level. And so on.

Taking all eight pairs of programmes shown on different days, the discrepancies were all 1 rating point or less, with no systematic bias towards over- or underprediction. The sequence of discrepancies (observed, minus predicted duplication) was

$$0, -1, -1, 1, 1, 0, 0, 0$$

Thus the average discrepancy was zero and there was certainly no general tendency for viewing of different pairs of programmes to be higher than predicted by the general law.

This result illustrates a practical application of the kind of research findings to be reported here. Analyses of past viewing patterns over thousands of programmes have provided us with a result (the duplication law) which now generally enables us to predict successfully the proportion of people who will see any two programmes. Hence we can test a particular issue of interest, e.g. whether programmes of a certain type attract an exceptionally large number of viewers in common.

With the "Arts Documentaries" there is no such tendency. The observed duplications were virtually the same as those for *any* kinds of programmes. People who watched one arts programme had no more tendency to see some other arts programme than to see, say, a western, or a religious programme, or a sports programme with a comparable rating. The pattern of viewing was therefore not related to programme content. We shall be discussing this type of result and its implications in more detail in later chapters.

Repeat-viewing

A pattern different from the duplication of viewing law occurs for repeat-viewing, i.e. the extent to which the same people view different episodes of the same programme.

In most cases, the size of the audience for successive episodes tends to be much the same. Table 2.2 shows this for four programmes (including news) which were shown on the ABC network over all five week-days in New York City in a typical week in 1974.

The ratings are clearly much the same on the different days, with Wednesday just fractionally higher. About 3 or 4% of all housewives watch "Love American Style" or "The 4.30 Movie" each day, about 11% watch the 6 pm "Eyewitness" news, and 8% the main "ABC News".

This steadiness of the ratings day by day might suggest that it is very much the same people watching at a given time each day. But this is largely not so – on average only about *half* the audience on one day also watch the same programme on another.

15

Table 2.2

Ratings for repetitive programmes on five week-days

New York Housewives Jan-Feb 74	% HW's Viewing					
	Mon	Tue	Wed	Thu	Fri	Av.
ABC						
4.00 Love Amer.Style	3	2	4	3	3	3
4.30 4.30 Movie	4	4	4	4	4	4
6.00 Eyewitness News 6	11	10	13	12	11	11
7.00 ABC Evening News	8	8	10	8	8	8
Average	7	7	8	7	7	7

Table 2.3 illustrates this in terms of the percentage of the Monday–Thursday audiences for each programme who also watched the *Friday* episode. Most of the figures are only in the 50's or 40's, with repeat-viewing in "The 4.30 Movie" slot even lower.

Table 2.3

Repeat-viewing within the week
(% of Monday to Thursday audiences
also watching on Friday)

New York Housewives Jan-Feb 74	% Viewing on Friday of the Audience on				
	Mon	Tue	Wed	Thu	Av.
ABC					
4.00 Love Amer.Style	38	47	39	56	45
4.30 4.30 Movie	29	26	43	51	37
6.00 Eyewitness News 6	59	55	56	58	57
7.00 ABC Evening News	52	50	43	49	48
Average	45	44	45	57	47

Most television programming is repetitive week by week, and here also audience sizes for successive episodes tend to be fairly steady. For example, roughly 15% might watch "Panorama" and about 25% "Star Trek" each week. Despite this relative steadiness of the ratings, the facts show that the general level of repeat-viewing in the UK is also only about 55% or so.

In analysing the repeat-viewing of different episodes of a programme, one may be dealing with a *serial* (i.e. a programme with a continuing story-line), or a *series* (i.e. the same characters appearing in episodes with self-contained plots, or comedy or variety shows with the same star performers). Less continuity occurs with drama slots ("The Monday Play"), screenings of old films ("The Tuesday Film"), regular current affairs programmes ("Midweek"), or the news. These various programme factors might be expected to influence the extent to which people view a programme regularly. But the findings show that in fact they appear to matter little. In general, repeat-viewing levels for different types of programme all tend to be about 55%. More detailed results are discussed in Chapter 5.

Rating levels

The outcome of any study of viewing patterns and audience flow might be expected to improve our understanding of why some programmes achieve higher ratings than others, but at present this is not so. If anything, the reverse is true. Thus in examining audience duplication for different programmes, the primary factors are usually the rating levels themselves. Given the ratings of two programmes we can successfully predict what proportion of the audience is in common.

In general, not much is known by way of systematic and generalisable results about the factors which make one programme of a certain type more popular than another of that type. There are apparently no well-established empirical findings or theoretical models from which ratings could be successfully predicted, nor can they even be systematically unravelled with hindsight. Predicting audience appeal and rating level is an area which is still very largely a matter of judgement.

Some statistical considerations

The results discussed in this book are all derived from *sample* data. Many of the illustrations are based on relatively small panels of, for example, 350 or so housewives in London. With samples of this kind, statistical sampling errors will occur in any particular case. These show themselves in terms of *irregularities* in the results.

However, the main findings reported here are essentially *regularities*. These could not arise as sampling errors since the regularities have generally been found to hold for a great variety of *different* samples, e.g.

housewives and men, people in different regions and data for different years. There is therefore no question of having to worry about the statistical significance of the main results, such as the duplication of viewing law itself or the relatively steady "55%" repeat-viewing levels that are reported.

Problems of statistical significance arise only with *discrepancies* from these regularities. Then the question is whether any particular discrepancy is mainly due to a "sampling error" (i.e. it would not have occurred in other similar samples), or whether it would also have shown up in other different samples or in a much larger sample.

Such problems are dealt with as they arise in the text. But as a general guide we can say that although many of the deviations in the illustrative tables are statistically not significant, little would be lost by regarding them as being real (i.e. statistically significant) and then asking if they have any *practical* significance.

Most of the deviations in question are relatively small. The crucial question is whether in some cases the deviations can themselves be seen to form some generalisable pattern. When they do (e.g. the deviations from the duplication of viewing law for late-evening programmes as mentioned earlier), such systematic and generalisable features are of course highlighted in the discussion. The point of interest then is not merely whether or not an isolated discrepancy is statistically significant but that the same kind of discrepancy occurs over and over again.

For ease of reading, statistical technicalities have been kept to a minimum in the text. The general statistical methodology for this kind of work has been discussed elsewhere (e.g. Ehrenberg, 1975).

The definition of a rating

To provide continuity in the exposition, many of the illustrations in this book are based on a sample of housewives in London (housewives are a convenient analysis group because by definition there is only one principal housewife per household). The resulting data then refer to all housewives in households in the London ITV area with sets capable of receiving the relevant transmissions (ITV and BBC1). That is the "population" in question in such a case. (We note that *guest* viewing is generally ignored here, since the measurements do not show whether guests on different days or in different weeks are the same people.)

Despite the emphasis on housewives in London, other population groups (e.g. men and "other women", i.e. non-housewives) have also been covered in the analyses, as well as other UK regions and data for the US. The findings reported here are therefore highly generalisable.

18

In discussing the data, we often talk about the "rating of a programme" or the "viewers of a programme", but the data strictly refer to quarter-hour time-bands (as described under Audience measurement in Chapter 1), and not to the whole programme. People usually watch a whole programme. In the case of half-hour programmes, about 95% of those who watch the first quarter-hour also watch the second. With much longer programmes more substantial erosion of the audience occurs – up to about 20% of initial viewers may be lost by the end, and even more late in the evening. Thus more attention will have to be paid to the precise definition of the viewing audience if one is literally concerned with the programme as an entity. But in describing the general run of results, we refer to the quarter-hour rating as if it related to the programme as a whole, for the sake of convenience. This does not affect any of the numerical results, only their precise meaning. For example, the ratings and repeat-viewing percentages for the ABC "Evening News" in Tables 2.1 and 2.2 refer to the first quarter-hour, 7 to 7.15 pm on Mondays, Tuesdays. etc. and not to the whole half-hour of the programme (plus commericals). The numerical results for the second quarter-hour would usually be marginally different (e.g. a Monday rating of 9 instead of 8 and a repeat-viewing percentage of 56 instead of 52), but the conclusions would not be affected.

Summary

In this chapter we have introduced the notion of audience flow – the extent to which the same audience watches different television broadcasts. There are two main results. One establishes that generally only about 55% of the viewers of one episode of a programme watch the next episode of that programme. The second, the "duplication of viewing law", states that the size of the audience common to two different programmes on different days depends on the ratings of the programmes and the channels on which they are shown, rather than the content of the programmes.

We now proceed to a more detailed discussion of these findings in the following chapters.

MATHEMATICAL APPENDIX: THE DUPLICATION OF VIEWING LAW

For the mathematically inclined, it is helpful to state the duplication of viewing law symbolically. For two times (or programmes) s and t, the law in its symmetrical form reads

$$r_{st} = r_s r_t k$$

Here r_{st} is the proportion of the audience watching both at times s and t, r_s and r_t are the audience ratings at these times, and k is a coefficient whose value is generally the same for different pairs of programmes.

Separate values of the coefficient k can be calculated for each pair of programmes, by the ratio $r_{st}/r_s r_t$ of the observed values of r_{st}, r_s and r_t. If the resulting values of k for different cases are all more or less the same, then the law with a *constant* value of k does in fact hold across this range of cases. This single value of k can then be estimated as the average of all the individual values of k.

A statistically more robust estimate of k is usually derived by forming the ratio of the total of the duplicated audiences for all pairs of times in question (i.e. Sum r_{st}) to the total of the cross-products $r_s r_t$ for all pairs of times (i.e. Sum $r_s r_t$), namely

$$k = \text{Sum } (r_{st})/\text{Sum } (r_s r_t)$$

The above version of the law is useful when analysing large amounts of data. The alternative formulation of the law mentioned earlier in this chapter consists of dividing the equation $r_{st} = r_s r_t k$ by r_t, so that it reads

$$\frac{r_{st}}{r_t} = r_s k$$

This version is helpful in allowing one to see, or demonstrate, the pattern in the data. Thus the ratio r_{st}/r_t is the proportion of the audience at time t who also watch at time s. The law says that this proportion should depend only on the rating at time s, as is discussed in the main text. This can be expressed in the language of "conditional probabilities" as $r_{s|t}$, i.e. as the probability of watching at time s given that one was watching at time t. We therefore have

$$r_{s|t} = r_s k$$

where the right-hand value does not depend on t.

If all the ratings are expressed as *percentages* rather than as proportions, the symmetrical form of the duplication law has to be written as

$$r_{st} = r_s r_t k/100$$

since each r value is multiplied by 100. But in percentage terms we still have

$$r_{s|t} = r_s k$$

The estimate of k for an individual pair of time-slots s and t is $k = (r_{st} \times 100)/r_s r_t$.

Example

If the observed value of r_{st}, the audience common to s and t, is 7·5% and the ratings at s and t are 25 and 15, then the estimated value of k is (7·5 × 100)/(25 × 15) = 2·0. If for other pairs of times the value of k is also about 2, the duplication law is operating with $k = 2·0$.

Conversely, if we already know that k is generally about 2·0 and we have ratings $r_s = 25$ and $r_t = 15$, then the predicted duplicated audience is

$$25 \times 15 \times 2·0/100$$
$$= 7·5$$

In other words, 7·5% of the population should watch *both* programmes.

As a proportion of the audience at time t, this is 7·5/15 = ·50 or 50%, which equals 25 × 2·0. Thus·50% of the audience at time t watch at time s. This is twice as high ($k = 2·0$) as the proportion of the total population who watch at s ($r_s = 25\%$).

Similarly, the proportion of the audience of s who watch at time t is 7·5/25 = 30%. The number of viewers at time t who also watch at time s is the same as the number of viewers at s who also watch at t (indeed, they are the same people). But this duplicated audience is a greater proportion of the smaller audience at time t (50%) than of the larger audience at time s (30%). This is in line with the ratio of the two ratings, $r_s = 25\%$ and $r_t = 15\%$.

3 Channel Loyalty

Channel loyalty is a major feature of audience behaviour. The term is however sometimes misunderstood. It does not imply that there are substantial numbers of people who view only one channel, but rather that individual viewers tend to show some degree of consistent preference for one channel over another. It is this phenomenon which can be summarised by the phrase "channel loyalty".

The basic question is, "How likely are viewers of one programme to watch another programme on the same channel rather than a programme on a different channel?" A pattern emerges within the context of the duplication of viewing law which can be interpreted as a preference or loyalty towards a channel. The argument involved is not complex but it is rather detailed. We therefore first summarise the main steps.

The main steps

People who saw a particular ITV programme yesterday are more likely than those who did not to watch a given ITV programme to-day. This tendency does not depend on the specific programmes or days in question but holds more generally; it also occurs for programmes on non-consecutive days and for ones in different weeks. Much the same pattern occurs on BBC channels, where viewers of one BBC programme are more likely than its non-viewers to watch a BBC programme on another day.

There is however no such positive tendency for audience flow *across* channels. Viewers of an ITV programme are if anything slightly *less* likely than its non-viewers to watch a given BBC programme. Similarly, viewers of a BBC programme are less likely than its non-viewers to watch a programme on ITV.

These findings therefore point to a degree of channel loyalty. Viewers of one channel are more likely to view that channel again than another channel. The strength of the preference varies little for the three channels in Britain, but with marginally more loyalty for the BBC. A similar pattern occurs in the US for the three national networks (as is discussed in Chapter 7), but with slightly less loyalty to each particular channel than in the UK.

The detailed results hang together quantitatively in terms of the duplication of viewing law introduced in the last chapter. The proportionality coefficient of this law reflects the extent of channel loyalty. Some numerical examples now serve to illustrate the detailed nature of channel loyalty.

Two ITV programmes

We start with two specific ITV programme transmissions in London in April 1971. They were: "The Mind of J.G. Reeder" (9 pm, Monday, 19 April) and "The Adventurers" (8 pm, Tuesday, 20 April). These two programmes had similar ratings amongst London housewives: 30% of the housewives saw "The Mind of J.G. Reeder" and 28% saw "The Adventurers".

The proportion of the population the two programmes had in common is established by cross-tabulating the viewing data. This shows that 14% of London housewives saw *both* of the programmes:

% of housewives seeing "J.G. Reeder"	30
"The Adventurers"	28
both	14

This means that 46 to 50% of the audience of one programme (14 out of 30 or out of 28) also saw the other. This level of audience overlap is high compared with the ratings of either programme. Thus Table 3.1 illustrates that "The Adventurers" was seen by 46% of the viewers of "J.G. Reeder", compared with only 20% of the *non-viewers* of "J.G. Reeder", and 28% of all London housewives. (The latter figure – the rating – lies closer to the percentage among the non-viewers of "J.G. Reeder" because there are more non-viewers of "J.G. Reeder", 70%, than viewers, 30%.)

Table 3.1

The audience overlap between the two programmes

London Housewives April 1971	% who viewed The Adventurers on Tuesday
Viewers of J.G. Reeder on Monday Non-Viewers of J.G. Reeder	46 20
All Housewives (the rating)	28

We can therefore conclude that there was a special tendency for viewers of the Monday ITV programme to view the one on Tuesday as well. Table 3.1 shows that they were more than twice as likely as the non-viewers of the Monday programme to do so.

Other ITV programmes

The question now is whether this high duplication of the two audiences reflects a general pattern or only something peculiar to the two programmes, or the times at which they were shown, or to some other specific factors. Both "J.G. Reeder" and "The Adventurers" were episodes in fictional series. The high audience duplication between them could therefore simply have been due to a special inclination amongst certain people to watch programmes of this particular type.

However, Table 3.2 shows that "The Adventurers" on Tuesday was also watched by between 40 to 50% of the housewife audiences of other "peak-time" ITV programmes on Monday evening. Thus "The Adventurers" was just about as popular with Monday viewers of "World in Action" (48%) at 8 pm and of "News at Ten" (43%) at 10 pm, as with viewers of lighter programmes such as "Opportunity Knocks" (50%) at 7 pm and "J.G. Reeder" itself (46%) at 9 pm.

Table 3.2

The audience overlap between various Monday programmes on ITV and "The Adventurers"

London Housewives April 1971			% who viewed The Adventurers Tuesday 8 pm
Viewers of			
Opportunity Knocks	- Monday	7 pm	50
World in Action	- "	8 pm	48
J.G. Reeder	- "	9 pm	46
News at Ten	- "	10 pm	43
All Housewives (the rating)			28

It therefore appears that the high audience duplication observed between "J.G. Reeder" and "The Adventurers" was not a function of any similarity between these two programmes as such, nor of the particular times at which they were shown. Instead, it seems to reflect a *general* tendency for the viewers of any ITV peak-time programme on Monday to have watched "The Adventurers" on Tuesday: 40 to 50% did so, compared with only 28% of the housewife population as a whole.

We now have to check whether this high audience overlap was specific to "The Adventurers" or also applied to other peak-time ITV programmes on Tuesday. Table 3.3 shows that other ITV peak-time programmes on Tuesday were also watched by at least 40% of the audiences of the various Monday programmes. Thus "Bless This House" at 7 pm on Tuesday was watched by more than 50% of any of the four previous evening's audiences, "The Saint" at 9 pm on Tuesday by between 40 and 50%, and so on.

Table 3.3

High overlap between Monday and Tuesday ITV programmes

London Housewives April 1971		% of Monday Audience who on TUESDAY Viewed			
		Bless this House 7 pm	The Adventurers 8 pm	The Saint 9 pm	News at Ten 10 pm
MONDAY					
Opportunity Knocks	- 7 pm	69	50	41	45
World in Action	- 8 pm	60	48	43	44
J.G. Reeder	- 9 pm	53	46	46	46
News at Ten	- 10 pm	53	43	39	53
Average		59	47	42	47
All Housewives (the rating)		33	28	23	29

All these overlap figures are high compared with the ratings of 23 to 33% of the Tuesday programmes themselves, shown at the bottom of Table 3.3. Thus markedly higher proportions of the viewers of any of the Monday ITV programmes watched each of the Tuesday ITV programmes.

The duplication of viewing law

The column averages in Table 3.3 vary from a high of 59 for "Bless This House" (7 pm Tuesday) to a low of 42 for "The Saint" (9 pm Tuesday). This variation in the overlap audiences largely mirrors the variation in the ratings themselves, which range from 33 for "Bless This House" down to 23 for "The Saint". The audience overlap or duplication tends to vary proportionally with the rating of the Tuesday programme. Table 3.4

26

shows this relationship more graphically by arranging the four Tuesday programmes in descending order of their ratings.

Table 3.4

Tuesday programmes in order of their ratings

London Housewives April 1971	% of Monday Audience who on TUESDAY Viewed				
	Bless this House	News at Ten	The Ad- venturers	The Saint	Ave- rage
Av. Duplication (T3.3)	59	47	49	42	49
Housewife Rating	33	29	28	23	28
Av. Dupl. /Rating	1.8	1.6	1.7	1.8	1.7

The ratios of the average duplication to the rating are very similar, ranging only from 1·6 to 1·8 and averaging about 1·7. We therefore can say that to a quite close degree of approximation the percentage of the viewers of a Monday ITV peak-time programme who watched a Tuesday ITV programme is about 1·7 times the percentage of the total housewife population who watched that programme. That is, the duplicated audience is generally higher than the rating by about 70% of the latter.

Table 3.5 compares these theoretical estimates (1·7 × rating) with the average observed duplication levels for each programme.

Table 3.5

The theoretical estimates 1·7 × rating

London Housewives April 1971	% of Monday Audience who on TUESDAY Viewed				
	Bless this House	News at Ten	The Ad- venturers	The Saint	Ave- rage
Average Duplication	59	47	49	42	49
1.7 x Rating	56	49	48	39	48

The agreement is clearly close, generally within a few percentage points. This is a case of the duplication of viewing law described in the last chapter. It has been found to occur in many thousands of cases, as will be illustrated further as we go along in this book.

The theoretical estimates of 1·7 × rating do not, however, agree quite as closely with the observed audience duplication between the *individual* Monday and Tuesday programmes shown in Table 3.3. There are

27

additional factors involved in these duplication patterns (including sampling errors, since the data are based on samples in the hundreds rather than the thousands). The largest discrepancy is for the overlap between the two 7 pm programmes, where the observed duplication is 13 percentage points higher than predicted. This is a systematic feature to which we shall return. Otherwise the discrepancies are on average only about 3 percentage points. Thus we can say that the *main* pattern in the data shows that the duplication levels are about 70% higher than the ratings.

Viewing on non-consecutive days

This high level of audience overlap between Monday and Tuesday programmes might occur because we have analysed viewing on *consecutive* days. It could be that TV sets tend simply to be kept tuned to the channel last watched the night before. However, further analyses show that the pattern illustrated so far, and the duplication of viewing law in particular, also holds for audiences on *non-consecutive* days.

Table 3.6 illustrates this for the Monday programmes and ones on Wednesday, two days later. Once more, the percentage of Monday ITV viewers watching an ITV programme on Wednesday is about 70% higher than the latter's rating (i.e. the average duplication equals 1·7 times the Wednesday rating). For example, 35% of all housewives watched "This is Your Life" at 7 pm on Wednesday (the rating), but about 60% of the *Monday* audiences did so (with again an abnormally high value of 69% for the two 7 pm programmes, just as in Table 3.3).

Table 3.6

Duplication of viewing between Monday and Wednesday ITV programmes

London Housewives April 1971		% of Monday Audience who on WEDNESDAY Viewed			
		This is Your Life 7 pm	I Spy 8 pm	Hine 9 pm	News at Ten 10 pm
MONDAY					
Opportunity Knocks	- 7 pm	69	60	53	50
World in Action	- 8 pm	62	62	50	49
J.G. Reeder	- 9 pm	55	57	50	49
News at Ten	- 10 pm	55	54	51	55
Average		60	58	51	51
1.7 x Rating		**60**	**61**	**54**	**53**
Rating		35	36	32	31

Table 3.7 illustrates a further result for the Monday programmes and ones on a Tuesday *two weeks later*. The pattern is much the same as before, but using a coefficient of 1·7 in the duplication law tends to overstate the observed results by a few percent. A coefficient of 1·6 would give a closer fit. It is not yet clear whether this suggestion of a slight erosion of duplication levels is general, since relatively little work has been done on such longer term viewing patterns.

Table 3.7

Duplication for the Monday ITV programmes and those on Tuesday
TWO WEEKS LATER

London Housewives April 1971		% of Monday Audience who Viewed on Tuesday (3 MAY)			
		Bless this House 7 pm	The Ad- venturers 8 pm	The Saint 9 pm	News at Ten 10 pm
Monday (19 APRIL)					
Opportunity Knocks	- 7 pm	51	55	42	40
World in Action	- 8 pm	45	50	43	41
J.G. Reeder	- 9 pm	46	50	46	42
News at Ten	- 10 pm	42	45	38	40
Average		46	50	42	41
1.7 x Rating		**44**	**54**	**48**	**48**
Rating		26	32	28	28

Channel switching

We have now seen that there is a relatively high overlap between the audiences of peak-time ITV programmes on different days. Viewers of one ITV programme are substantially more likely than non-viewers of that programme to watch another ITV programme.

This high duplication of viewing for ITV programmes need not by itself imply any form of "loyalty" to the ITV channel. For example, viewers of an ITV programme could also be more likely than non-viewers to watch a *BBC* programme.

The high duplication for the ITV programmes implies special "loyalty" to the ITV channel only if such high audience duplication does not occur *between* channels, e.g. between the audience of an ITV programme on the one hand and that of a BBC programme on the other.

In practice, between-channel audience overlap is in fact relatively low. This is illustrated in Table 3.8 for the same Monday and Tuesday that we have already been examining (19 and 20 April 1971). The data show the extent to which viewers of the ITV peak-time programmes on the Monday watched the BBC1 peak-time programmes the next day.

Table 3.8

ITV versus BBC1: the tendency for Monday's ITV audiences
to view BBC programmes on Tuesday

London Housewives April 1971	% of Monday ITV Audience who on Tuesday viewed BBC1's			
	Top of the Form 7 pm	A Shot in the Dark 8 pm	News 9 pm	Civilis- ation 10 pm
Monday ITV				
Opportunity Knocks – 7 pm	12	23	24	7
World in Action – 8 pm	12	24	26	6
J.G. Reeder – 9 pm	12	23	28	9
News at Ten – 10 pm	9	25	29	9
Average	11	24	27	8
0.8 x Rating	**15**	**22**	**25**	**9**
Rating	19	28	31	11

Thus "Top of the Form" at 7 pm Tuesday on BBC1 was watched by between 9 and 12% of the viewers of any one of the Monday ITV programmes; "A Shot in the Dark" at 8 pm on BBC1 was watched by about 24% of the Monday ITV viewers, and so on. These duplication levels are lower than the ratings of the BBC programmes themselves. As shown in the bottom line of Table 3.8, about 19% of London housewives watched "Top of the Form" at 7 pm whereas only 9 to 12% of the ITV viewers did so. The same differences occur for the later BBC programmes. *Fewer* ITV viewers therefore watched the BBC1 programmes than the proportion watching in the population as a whole.

The pattern still follows the duplication of viewing law, within limits of a few percentage points. But the proportionality factor is very different. It is 0·8 for ITV and BBC1 as compared with 1·7 for two ITV programmes (Table 3.4). A duplication coefficient of 0·8 means that viewers of the ITV programmes are 20% *less* likely to watch the BBC1 programmes than the population as a whole. Viewing of an ITV programme therefore appears to *inhibit* the viewing of BBC1 programmes on another day.

The same pattern holds the other way round. Viewers of BBC1 programmes are less likely than the general population to switch to ITV programmes on another day. This is illustrated in Table 3.9 for the same programme combinations as in Table 3.8.

Table 3.9

BBC1 versus ITV: the tendency for BBC1 audiences on Tuesday to view ITV programmes on Monday

London Housewives April 1971		% of Tuesday BBC1 Audience who on Monday viewed ITV's			
		Opportunity Knocks 7 pm	World in Action 8 pm	J.G. Reeder 9 pm	News at Ten 10 pm
Tuesday BBC 1					
Top of the Form	- 7 pm	20	18	20	15
A Shot in the Dark	- 8 pm	26	26	25	29
News	- 9 pm	24	25	27	29
Civilisation	- 10 pm	21	15	23	26
Average		23	21	24	25
0. 8 x Rating		**26**	**24**	**24**	**26**
Rating		32	30	30	32

We therefore have a general pattern that viewers of an ITV programme on one day, when faced with the choice between similarly rated ITV and BBC1 programmes on another day, are about *twice as likely* to watch an ITV programme again as to tune to BBC1 that day (the duplication coefficients of 1·7 and 0·8 respectively). This is why the high duplication level between ITV programmes is in fact a reflection of channel loyalty.

High duplication on BBC

So far our analysis of audience overlap for programmes shown on the same channel has been restricted to ITV, but the results for BBC programmes are similar. The general pattern is illustrated in Table 3.10 for programmes shown on the same Monday and Tuesday.

There is some variability in the individual results, with an exceptionally high value occurring once more for the two 7 pm programmes. But it is clear that viewers of a Monday BBC1 programme are substantially more likely to watch a Tuesday BBC1 programme than is the population as a

31

Table 3.10

High overlap on BBC1

London Housewives April 1971	% of Monday BBC1 Audience who on TUESDAY Viewed BBC1's			
	Top of the Form 7 pm	A Shot in the Dark 8 pm	News 9 pm	Civilisation 10 pm
MONDAY BBC1				
A Question of News - 7 pm	49	51	49	28
Panorama - 8 pm	46	48	56	23
News - 9 pm	39	45	55	28
Brett - 10 pm	31	43	47	21
Average	41	47	52	25
1.8 x Rating	34	50	56	20
Rating	19	28	31	11

whole. The duplication levels are generally almost twice as high as the programme ratings, averaging at a factor of 1·8. This is typical of the more general run of results for BBC1.

The duplication of viewing law therefore also applies to BBC1 programmes, and with a similar degree of overlap as for ITV — the coefficient in April 1971 was 1·8 for BBC1 programmes, compared with 1·7 for ITV programmes.

Duplication on BBC2

BBC2 programmes generally attract smaller audiences than programmes on the other two channels. These lower rating levels might imply some special pattern of viewing, e.g. viewers being more "devoted" to the channel, or perhaps especially selective in what they view.

Viewing of BBC2 transmissions also follows the duplication of viewing law but the apparent degree of audience overlap is much higher than on the other two channels. Table 3.11 illustrates overlap between the audiences of BBC2 programmes on different days (shown for the average pair of week-days, because the sample sizes of homes able to receive BBC2 in April 1971 were small). It shows that housewife viewers of a BBC2 programme on one day are about 3 times as likely to watch a BBC2 programme on another day as is the general population of housewives.

However, this high duplication is more apparent than real. There is a simple explanation. In 1971 a substantial number of housewives virtually

Table 3.11

High duplication on BBC2
(The average for all pairs of week-days)

London Housewives April 1971	% of BBC2 Audience who on Another Day Viewed BBC2 at:		
	8 pm*	9 pm	10 pm
BBC2			
Average Week-day - 8 pm*	12	14	31
" " - 9 pm	11	17	25
" " - 10 pm	12	16	21
Average	12	16	26
3.2 x Rating	13	16	26
Rating	4	5	8

* There were no 7 pm BBC2 transmissions

never watched BBC2 because they did not have a set capable of receiving transmissions on this channel. Thus the duplication coefficient of 3·2 is high not because channel loyalty is specially marked, but because the BBC2 *ratings* were depressed by this non-availability factor. Indeed, the duplication levels of about 10 to 30% shown in Tables 3.11 are not high in absolute terms, but only relative to the very low programme ratings.

The ratings of BBC2 programmes are therefore adjusted in Table 3.12 for "availability", i.e. the percentage of housewives viewing the programme amongst those with sets able to receive BBC2 (about 60% of London homes in 1971). This reveals a pattern of audience overlap much more like that for ITV and BBC1. Compared with the higher adjusted ratings among homes with BBC2 sets, the duplication coefficient is now about 1·9, only slightly higher than the coefficients for the two main

Table 3.12

BBC2 duplication compared with the ratings in BBC2 homes

London Housewives (with sets capable of receiving BBC2)	% of BBC2 Audience who on Another Day Viewed BBC2 at:		
	8 pm	9 pm	10 pm
Av. duplication (T.3.11)	12	16	26
1.9 x Rating	13	15	25
Rating (in BBC2 Homes)	7	8	13

channels (1·7 and 1·8). The general pattern of channel loyalty in 1971 was thus broadly the same for all three channels.

Switching to BBC2

One of the arguments used in 1964 to support the award of the third TV channel to the BBC was that this would permit complementary rather than competitive programming. Instead of competing directly with ITV for audiences and hence providing similar "popular" programmes, BBC2 was to offer a real alternative, providing viewers with a greater choice and satisfying more minority interests. With such a complementary programming strategy one might expect to find a special link between BBC1 and BBC2.

Table 3.13 illustrates the duplication pattern between BBC1 and BBC2 for the average pair of week-days in the week of 19 April. It shows that the viewer of a BBC1 programme on one day tends to be fractionally more likely to watch BBC2 broadcasts on another day than is the average housewife with a BBC2-type set. The duplication coefficient is 1·1.

Table 3.13

Duplication between BBC1 and BBC2
(the average for all pairs of week-days)

London Housewives April 1971	% of BBC1 Audience who on Another Day Viewed BBC2 at:		
	8 pm	9 pm	10 pm
BBC 1			
Average Week-day - 7 pm	11	11	12
" " - 8 pm	9	9	13
" " - 9 pm	9	8	12
" " - 10 pm	8	7	15
Average	9	9	13
1.1 x Rating	8	9	**14**
Rating in BBC2 Homes	7	8	13

The comparable figures for ITV viewers watching BBC2 on another day are set out in Table 3.14. Here the tendency is for an ITV viewer to be slightly *less* likely to watch a BBC2 programme than is the average housewife. The duplication coefficient is 0·9.

34

Table 3.14

Duplication between ITV and BBC2
(the average for all pairs of week-days)

London Housewives April 1971	% of ITV Audience who on Another Day Viewed BBC2 at:		
	8 pm	9 pm	10 pm
ITV			
Average Week-day - 7 pm	6	7	12
,, ,, - 8 pm	6	7	10
,, ,, - 9 pm	6	8	14
,, ,, - 10 pm	5	7	11
Average	6	7	12
0.9 x Rating	6	7	12
Rating in BBC2 Homes	7	8	13

The important aspect of these results is not the difference between the two duplication coefficients of 1·1 and 0·9, but their similarity. To a first approximation both coefficients are 1. Furthermore, they are not very different from the duplication level for BBC1 and ITV programmes seen earlier (a coefficient of 0·8).

Roughly speaking, viewers of either a typical BBC1 programme or a typical ITV programme are therefore just about as likely to watch a BBC2 programme on another day as is the average housewife with a BBC2 set. BBC1 viewers tend to be about 10% *more* likely to do so and ITV viewers about 10% *less* likely. These differences therefore point to only a small degree of overall loyalty between the two BBC channels. But the link between BBC1 and BBC2 is far less marked than the loyalty towards either of the individual BBC channels or towards ITV.

Summary

Table 3.15 gives the constants of the duplication of viewing law for April 1971, as they have been developed in this chapter and recurred among thousands of different programme combinations at that time.

The within-channel constants, all substantially above 1, show that channel loyalty exists for viewing on different days of the week (and across different weeks). BBC2 and BBC1 appear to have attracted a slightly more loyal following, but the similarity of the various duplication

Table 3.15

Constants of the duplication of viewing law in 1971

Within-channel duplication constants :		
	ITV x ITV	= 1.7
	BBC1 x BBC1	= 1.8
	BBC2 x BBC2	= 1.9
Between-channel duplication constants :		
	ITV x BBC1	= 0.8
	ITV x BBC2	= 0.9
	BBC1 x BBC2	= 1.1

constants is more remarkable than the marginal differences. The between-channel constants also show some slight BBC link, but again the similarities and closeness to a value of 1 are the striking points.

To summarise, in 1971 a peak-time programme with a rating of 20 was generally watched by just under 20% of the audience of any peak-time programme shown on a *different* channel on another day of the week, but by roughly 34% of the audience for another day's peak-time programme on the *same* channel.

4 Other Factors

The results on channel loyalty and audience duplication described so far were mainly restricted to viewing on week-days in "peak-time" (about 7 to 10.30 pm) for London housewives in April 1971. To gain a more complete picture of audience behaviour, results under a wider range of conditions need to be considered.

In this chapter we first consider some simple generalisations across different demographic groupings (e.g. men and women; geographic regions), together with trends over the years. Secondly, we discuss more complex situations, namely week-end viewing, off-peak viewing, and duplication of viewing between programmes shown on the same day — the "inheritance effect". Thirdly, we examine the influence of different types of programmes. We conclude with a specific application of the findings to a change in the programming policy for news broadcasts.

Demographic factors

In general, the duplication of viewing law holds for different demographic groups. Table 4.1 compares the duplication coefficients for London housewives discussed in the last chapter with those for London men and

Table 4.1

Summary of duplication coefficients for housewives and men

April 1971	Duplication-coefficients for:		
	London Housewives	London Men	Lancashire Housewives
Within ITV	1.7	1.8	1.6
Within BBC1	1.8	1.8	1.6
Within BBC2	1.9	1.8	(1.4)*
Between ITV and BBC1	0.8	0.9	0.9
Between ITV and BBC2	0.9	0.9	(1.0)*
Between BBC1 and BBC2	1.1	1.2	(1.2)*

* Small sample base

37

Lancashire housewives in April 1971. The results are very similar. The within-channel duplications in Lancashire are fractionally lower than in London and there are other small differences, but the broad pattern of results is the same. This basic pattern has also been found to hold across the whole of the UK.

Changes over time

The results described so far are not new. The same sort of duplication patterns have now been noted over a period of years. However, the numerical values of the duplication coefficients changed in about 1968, as is summarised in Table 4.2. (Results for 1974, not available at the time of writing, will be given in Appendix B.)

Table 4.2

Summary of duplication coefficients in the UK, 1966–71

Average Results	Within-Channel (ITV OR BBC 1)	Between-Channel (ITV AND BBC 1)
1966	1.3	1.0
1967	1.4	1.0
1969	1.7	0.8
1971	1.7	0.8

In the late sixties and early seventies, the within-channel duplication coefficient in the UK has been about 1·7 and that between ITV and BBC1 about 0·8, the same as for the specific data for April 1971 discussed in Chapter 3. Viewers of an ITV programme were therefore about 70% more likely than the average member of the potential viewing population to watch an ITV programme on another day, and 20% *less* likely to watch a BBC1 programme.

In the mid-sixties, however, the within-channel duplication coefficient had been generally about 1·4 and that between ITV and BBC1 about 1·0. In these earlier years, viewers of an ITV programme were therefore only about 40% more likely than the average member of the population to watch an ITV programme on another day, and *just* as likely as the latter to watch a BBC1 programme. Thus sometime between 1967 and 1969 there appears to have been a greater polarisation of television audiences in the UK.

During this period there were changes in the regional programme companies licensed by the IBA to broadcast ITV, suspension of ITV transmissions for several weeks during a strike at the point of changeover (and hence enforced viewing of the BBC), changes in the market research company commissioned to collect TV audience measurements for JICTAR, and some growth in the number of households owning sets capable of receiving BBC2 transmissions. Whatever the cause, the available measurements denote that viewers are now somewhat more likely to stick to their chosen channel and less likely to switch channel than appeared to be the case before. This greater polarisation of the audience is contrary to beliefs expressed in the last few years.

Viewing at off-peak times

Certain exceptions to the simple duplication of viewing law also occur. The major ones are in the afternoon and early evening on the one hand, and late in the evening on the other hand.

Audience duplication between afternoon and early-evening programmes shown on different week-days on a given channel is generally substantially higher than the audience overlap between programmes shown at peak-times. (This was already foreshadowed by the abnormally high duplications for pairs of programmes at 7 pm, as noted in the previous chapter.) High duplication also occurs for pairs of programmes shown *late* in the evening on different days.

Table 4.3 illustrates the findings for some late-afternoon programmes on ITV, for London housewives in April 1971. Audience overlap is substantially higher than indicated by the duplication law using the "peak-time" coefficient of 1·7. Of those London housewives who saw "Lost in Space" at 5 pm on Monday; 37% also watched ITV at 5 pm on Tuesday. This is more than 4 times as high as the 9% of all London housewives who watched then.

Again, 51% of the viewers of "Lost in Space" watched "Today" at 6 pm on Tuesday, which is 2½ times as high as its housewife rating of 20. Not only is the overlap generally higher than at peak-times, but the extent of it also varies from case to case (the overlap usually being highest at exactly the same times on the different days). This is the general finding for afternoon and early-evening viewing on week-days.

Similarly, analysis of late-night viewing in April 1971 shows that of

Table 4.3

High audience duplication in the late afternoon

London Housewives April 1971	% of Monday Audience who on TUESDAY Viewed ITV	
	Junior Showtime 5 pm	Today 6 pm
MONDAY ITV		
Lost in Space - 5 pm	37	51
Today - 6 pm	21	62
Average	25	54
1.7 x Rating	**15**	**34**
Rating	9	20

those who watched ITV at 11 pm on the Monday, 39% watched ITV again at the same time on Tuesday. This is 3 times higher than the 13% rating in the whole population.

The question now is in what ways these early or late off-peak viewers differ from peak-time viewers. First of all, there are fewer of them (the ratings are lower at these times, which is why they are called off-peak). But since so many off-peak viewers watch programmes at similar times on different days, it might appear that they are especially avid viewers of television.

This is not really so. Off-peak viewers do not differ from peak-time viewers in their viewing of *peak-time* programmes. This is illustrated in Table 4.4. The striking finding is that the early-evening viewers on Monday generally behaved just like the viewers of later programmes in terms of the proportion who watched the next day's peak-time pro-grammes. For example, "The Adventurers" on Tuesday was seen by roughly 50% of both the early-evening and of the peak-time viewers on Monday. The same was true for "The Saint" at 9 pm and for the "News at Ten". But "Bless This House" at 7 pm on Tuesday was on the border line of the "early evening" effect. It was viewed by a somewhat higher proportion (69%) of the Monday late-afternoon and early-evening viewers than of the peak-time viewers (55%). The pattern shown in Table 4.4 gen-eralises to other days and also to late-night viewers. These also hardly differ in their viewing of peak-time programmes on other days from the general patterns of peak-time viewers themselves.

Table 4.4

The duplication of early and peak-time ITV audiences

London Housewives April 1971		% of Monday Audience who on TUESDAY Viewed			
		Bless this House 7 pm	The Ad- venturers 8 pm	The Saint 9 pm	News at Ten 10 pm
MONDAY					
Lost in Space	- 5 pm	69	43	46	57
Today	- 6 pm	69	57	49	52
Opportunity Knocks	- 7 pm	69	50	41	45
World in Action	- 8 pm	60	48	43	44
J.G. Reeder	- 9 pm	53	46	46	46
News at Ten	- 10 pm	53	43	39	53
Average		62	48	44	49
1.7 x Rating		**56**	**48**	**39**	**49**
Rating		33	28	23	29

It follows that the duplication of viewing law as such holds and that off-peak viewers are not unusual in their peak-time viewing. Instead, they merely have a higher tendency than the population at large to view at similar off-peak times on other days. The explanation is that it is the *non-viewers* at off-peak times who are abnormal.

Many people are not at home on week-day afternoons or early evenings, or alternatively do not stay up late in the evening, e.g. past 10 or 11 pm. These things are generally regular social habits — working hours, or one's choice of bedtime, tend to be the same on different days of the week.

The high audience duplication at off-peak times therefore reflects the relatively small pool of available viewers at these times. The ratings are depressed because many people are not available. In comparison with these low ratings, the audience overlap for those viewers who *can* watch is then high.

This is similar to the explanation of the apparently high duplication on BBC2 noted in the last chapter (Table 3.11). This was also due to non-availability, in the sense that an appreciable proportion of people were not able to receive BBC2 in 1971. (For the analysis of off-peak viewing no direct quantitative data on the pattern of viewers' absences from home or their bedtime habits are available, so that no detailed numerical reconciliation of the observed results is yet possible.)

Week-end viewing

The above interpretation is supported by the findings for week-end viewing, when social habits tend to be different. Week-end viewing also follows the duplication of viewing law, as illustrated in Table 4.5 for the overlap between Monday peak-time audiences and those on the following *Sunday*. But the proportionality factor in this case was only 1·3 compared with the 1·7 value for the more general week-day results discussed earlier.

Table 4.5

Audience overlap between week-days and week-ends

London Housewives April 1971		% of Monday Audience who on SUNDAY Viewed			
		Stars on Sunday 7 pm	The Ceremony 8 pm	The Ceremony 9 pm	News 10 pm
MONDAY					
Opportunity Knocks	- 7 pm	39	62	64	55
World in Action	- 8 pm	32	61	62	52
J.G. Reeder	- 9 pm	26	52	60	57
News at Ten	- 10 pm	25	48	61	61
Average		30	56	62	56
1.3 x Rating		29	52	60	60
Rating		22	40	46	46

The relatively lower duplication indicates that on Sunday evening the ITV audience was joined by some viewers who were relatively light viewers of the channel during the week. This might be another feature due to general social habits — some people have more leisure to view at week-ends. But Table 4.5 applies to April 1971, and at other times the week-end effect has been less marked and the duplication coefficient is closer to that for pairs of week-days. This may be due to seasonal factors affecting social habits, but the general character of the situation is not yet fully established.

Table 4.6 extends the picture to week-end afternoon audiences. This reveals higher audience duplication between week-day and week-end afternoon audiences, the proportionality factor being 2·3. This higher duplication is however much less marked than the factors of 3 or 4 seen in Table 4.3 for off-peak duplication on two *week-days*. The reason is that at the week-end, the *potential* audience is greater than on week-days, thus lowering the degree of duplication from that which occurs on two week-days.

Table 4.6

AFTERNOON overlap between week-days and week-ends

London Housewives April 1971	% of Monday Audience who on SUNDAY Viewed			
	The Big Match 2 pm	Captain Boycott 4 pm	The Golden Shot 5 pm	H.R. Pufnstuf 6 pm
MONDAY				
Lost in Space - 5 pm	29	46	49	46
Today - 6 pm	21	40	47	35
Average	25	43	48	40
2.3 x Rating	25	46	48	39
Rating	11	20	21	17

The inheritance effect

We now turn to how audiences view different programmes shown on the same channel on the same evening. This has perhaps been one of the more misunderstood and casually interpreted aspects of audience measurement. Programme planners and television commentators have often subscribed to the belief that once viewers have switched on their sets, they tend to continue watching the same channel throughout the evening. Such a belief puts a high premium on attracting a large audience early in the evening. Once attracted, it is regarded as relatively easy for subsequent programmes on that channel to "inherit" part of this audience.

However, this common belief may not necessarily be correct. It has arisen because the within-channel audience overlap for two programmes on the same day is in fact high compared with the ratings of each programme. Table 4.7 illustrates this for peak-time ITV programmes on a Monday night. The ratings are in the low 30's, the duplication in the high 50's and 60's.

However we have already seen that within-channel levels of audience duplication tend to be higher than ratings *even if the programmes are transmitted on different days*. This is the channel loyalty phenomenon isolated in Chapter 3, but its existence has not been generally appreciated (especially as the BBC's one-day recall measurement system gives no

Table 4.7

High duplications on the same day

London Housewives April 1971		% of Monday Audience who on SAME Monday watched ITV:			
		Opportunity Knocks 7 pm	World in Action 8 pm	J.G. Reeder 9 pm	News at Ten 10 pm
MONDAY ITV					
Opportunity Knocks	- 7 pm	-	65	55	55
World in Action	- 8 pm	69	-	69	54
J.G. Reeder	- 9 pm	57	69	-	65
News at Ten	- 10 pm	55	51	62	-
Monday Rating		32	30	30	32

information about audience flow between different days). The results show that the high degree of audience duplication illustrated in Table 4.7 turns out to be virtually identical with that to be expected from channel loyalty generally, *except* for pairs of programmes that are shown very close to each other.

This is illustrated by Table 4.8, where predictions from the *between-day* duplication of viewing law are compared with the observed duplication levels between the different ITV programmes on the same day.

These predictions hold to within a few percentage points for the six programme pairs which are separated by more than an hour. The relevant deviations in Table 4.8 are only 4, 3, 1, 1, 0 and 0 (averaging 1·5), as shown in the bottom half of the table where the deviations between the observed and predicted figures are given. More generally, across a wide range of other cases, such deviations have tended to average at zero. The between-day law therefore successfully predicts the within-day duplication pattern for the more widely separated programmes. Generally there is *no* special overlap for programmes on the same day.

In contrast, the observed overlap is always consistently higher for adjacent or adjacent but one pairs of programmes. This is seen well from the bottom half of Table 4.8, the selection of illustrative programmes here being broadcast at hourly intervals. A higher proportion of the viewers of one ITV programme will watch the *next* ITV programme than if the two programmes were shown on different days altogether. Much the same effect occurs on the BBC channels.

Table 4.8

Observed same-day duplications (O), predictions (P)
and deviations (O−P)
(predictions based on the between-day duplication law
with coefficient 1·7)

London Housewives April 1971	% of Monday Audience who on SAME Monday watched ITV:			
	Opportunity Knocks" 7 pm	World in Action 8 pm	J.G. Reeder 9 pm	News at Ten 10 pm
MONDAY ITV	O P	O P	O P	O P
Opportunity Knocks - 7 pm		65 51	55 51	55 54
World in Action - 8 pm	69 54		69 51	54 54
J.G. Reeder - 9 pm	57 54	69 51		65 54
News at Ten - 10 pm	55 54	51 51	62 51	-
Deviations	(O-P)	(O-P)	(O-P)	(O-P)
Opportunity Knocks - 7 pm	-	14	4	1
World in Action - 8 pm	15	-	18	0
J.G. Reeder - 9 pm	3	18	-	11
News at Ten - 10 pm	1	0	11	-

The existence of such "inheritance" or "lead in" effects has been known for a long time. Kirsch and Banks (1962) for example noted high correlations in the USA among the audiences of programmes on the same network on the same evening, and found the highest figures among immediately adjacent programmes. But in the absence of norms for the *usual* amount of audience duplication, neither the size nor the limitations of this effect could be established (e.g. that the correlations for non-adjacent programmes are not *specially* high).

These then are general findings. It is now clear why many interested parties have overestimated the significance of catching a large audience early in the evening. It is because the basic nature of channel loyalty (as reflected in the duplication of viewing law) was not widely understood. One *could* say that there is a *general* inheritance effect for programmes on the same channel shown on different days (i.e. channel loyalty). The apparent pattern of inheritance on any single evening is then only a facet of this general effect. The *special* inheritance effect for programmes close together is additional to this and can be due to three possible causes. Either people stay tuned to the next programme out of inertia, or because the programme has ended part-way through the programmes on the the alternative channels, or because they tuned in to the *previous* programme to wait for the programme scheduled to appear next (a "lead out" rather than a "lead in" effect).

Programme types

Implicit in the findings described so far is the conclusion that while a programme's content undoubtedly affects the size of its audience or "rating", content has little or no additional effect on the degree of its audience duplication with any other programme. This is a somewhat startling result. Many people would expect that "clusters" of programmes exist which tend to be viewed by certain groups of viewers. For example, one might suppose that there are viewers who like situation comedies and hence view all or most of them. Alternatively, one might expect a negative, inhibitory effect, e.g. that being addicted to *one* situation comedy or *one* western series is enough — there is no need to watch the other series of the same type.

But in general, neither of these views fits the facts of actual viewing behaviour. Instead, we find that audience overlap for the population at large can be successfully predicted by the duplication of viewing law (plus systematic deviations for early-evening viewing, etc., which relate to audience availability). These predictions can be made without taking into account what particular programme or type of programme each audience is watching.

Any programme-type preferences can in effect exist only within the limits of the deviations of the observed duplication figures from the predictions of the duplication of viewing law. These limits have been illustrated in the various tables in this and the preceding chapters. The deviations are generally small, of the order of a few percentage points.

However, *some* scope for systematic patterns remains. To demonstrate this, the deviations from the simple duplication law are illustrated once more in Table 4.9 for the Monday and Tuesday ITV programmes discussed earlier. The largest discrepancy, +13 for the two 7 pm programmes, reflects the systematic late-afternoon/early-evening pattern which has already been noted. But the other deviations still range over 12 percentage points, from −5 to +7 (although the *average* size of the deviations is still only about 3 points). The question now is whether any generalisable patterns are reflected by the relatively low duplication between "World in Action" and Tuesday's "News at Ten" say, or by the high duplication between "J.G. Reeder" and "The Saint".

Table 4.9

Observed duplications (O), predictions (P) and deviations (O−P)
(from Tables 3.3 and 3.5)

	% of Monday Audience who on TUESDAY Viewed:			
London Housewives April 1971	Bless this House 7 pm	The Ad- venturers 8 pm	The Saint 9 pm	News at Ten 10 pm
MONDAY	O P	O P	O P	O P
Opportunity Knocks - 7 pm	69 56	50 48	41 39	45 49
World in Action - 8 pm	60 56	48 48	43 39	44 49
J. G. Reeder - 9 pm	53 56	46 48	46 39	46 49
News at Ten - 10 pm	53 56	43 48	39 39	53 49
Deviations	(O-P)	(O-P)	(O-P)	(O-P)
Opportunity Knocks - 7 pm	13	2	2	-4
World in Action - 8 pm	4	0	4	-5
J. G. Reeder - 9 pm	-3	-2	7	-3
News at Ten - 10 pm	-3	-5	0	4

Virtually no significant patterns in these deviations have however been found for any particular type of programme. Table 4.10 illustrates the first such analysis, for data for the first week in May 1967. It sets out twenty programme categories and subclassifications for ITV and BBC1 programmes (excluding children's programmes), together with the number of programmes in each category and their average ratings. The last column gives the average deviations of the observed audience duplication from the predictions of the duplication of viewing law for each pair of programmes within the category. In analysing such aggregated data it is more convenient to express the readings as percentages of the population seeing both programmes of a pair, rather than as percentages of the audience of one also seeing the other. (See also Appendix, Chapter 2 for this formulation of the duplication of viewing law.)

The average deviations are generally less than 1 rating point, which is small compared with the audience rating of each programme. In effect there are therefore virtually no systematic differences from the law's predictions and no evidence of clusters by programme type.

"Westerns" provide the only programme grouping for which the average deviation is more than 1 rating point. To this extent there was a special tendency for people to watch two or more of the westerns as a group. However, this result is based on only six programmes, is of doubtful statistical significance, has not been confirmed at other points in time, and

Table 4.10

Deviations of the observed duplications from the duplication of viewing law, analysed by twenty standard IBA programme categories and classes
(pairs of programmes on different evenings but excluding programme pairs near 6 pm; ITV & BBC1, 1 to 7 May 1967, London & North housewives)

CATEGORY & Class	Number of pro-grammes	Average rating	Av. observed minus theoretical duplications*
NEWS (Week-day)	40	23	-.3
NEWS MAGAZINES	22	21	.6
DOCUMENTARIES & NEWS FEATURES			
News Features	19	15	.4
General Discussion	11	15	.5
Magazines	4	16	.5
Arts	9	12	-.4
Miscellaneous	12	15	.1
RELIGION	20	7	.1
ADULT EDUCATION	7	5	.4
PLAYS	8	20	.7
DRAMA & SERIALS			
Series	9	26	-.2
Adventure & Crime	29	28	.1
Westerns	6	25	1.2
Serials	12	28	.0
CINEMA FILMS	14	32	-.1
ENTERT. & MUSIC			
Comedy Series	24	23	.1
Contests	12	26	.7
Light Music	9	19	.1
Other	11	22	.1
SPORT	25	15	.8
AVERAGE	15	21	.2

* Duplications expressed as percent of the population

is small anyway. It seems therefore at most the kind of (small) exception to prove the rule that as far as the IBA's common-sense programme classifications are concerned, there is no special tendency across the population for people who watch *one* programme of a given type also to watch others of the same type.

Checks among *regular* viewers of programmes (i.e. those who see at least three episodes out of four) have also shown no evidence so far of any clustering in viewing behaviour by programme type. Nor does clustering show itself in one-person households, where individual viewing preferences could operate irrespective of conflicting preferences of different family members.

Some writers in the US, e.g. Swanson (1967), Bruno (1973), and Gensch and Ranganathan (1974), have reported apparent programme-category effects derived from applying statistical techniques like factor analysis to certain types of viewing data. But Swanson, for example, did not take account of the choice of channel (or day of week) and failed to note that all the programmes in his first factor were transmitted by the ABC network, and all those in the second factor by CBS, and so on (Ehrenberg, 1968). The high correlations reflected by these so-called programme factors were therefore no more than a reflection of channel loyalty. The other studies cited are also open to interpretative difficulties, including differences in *reported* viewing behaviour (as discussed in Chapter 9) and directly *measured* behaviour.

A specific application: news broadcasts

A practical application of the general results outlined in this and the preceding chapter is illustrated by an evaluation of the audience for news broadcasts. This case arose some years ago when certain decisions had to be made regarding ITV's news bulletins. Until July 1967 both of the main TV channels in Britain carried two ten-minute news bulletins per day, screened on week-days at about 6 and 9 pm.

In July 1967 the later ITV bulletin was increased to thirty minutes to allow greater depth of treatment and was also moved from 9 to 10 pm. This involved *three* changes: the time of the bulletin, its length, and the fact that it was no longer opposite the 9 pm bulletin on BBC1. To evaluate the situation, studies were made in May 1967 (well before the change), just *after* the change in July, and in October when viewing behaviour appeared to have fully settled down again.

In trying to assess the public's response to the flow of news, one might have measured how well major news items had been communicated (in May 1967 these included violence in Aden, Britain making an application to join the European Common Market, and hopes for a quieter jet engine). But it was not clear how any assessment of communication (by currently feasible research techniques) could tell one much about the effectiveness of TV news bulletins in general, and the effect of the July programme change in particular.

A more direct, if limited, evaluation seemed to be in terms of people's actual *utilisation* of the news bulletins, namely in terms of their viewing behaviour. How many people did not see *any* news? At the other extreme, was there any special following for TV news, and if so, was this increased or decreased by the change to a later and longer ITV bulletin?

The ratings tell us the audience size achieved by each separate news bulletin, but a further question is what use people make of the *different* bulletins available to them. In what combinations do people view bulletins on the same day and on different days, on the same channel and on different channels, at 6 pm and at 9 or 10 pm? Given that the contents of the different bulletins on the same evening would usually not vary greatly, did viewing of one bulletin per evening suffice and therefore inhibit the watching of any others?

To try to answer such questions it is not enough merely to tabulate viewing patterns. One has also to be able to predict what such patterns would generally be like. Only if the factors involved are sufficiently well understood to permit successful prediction can the effect of variations in any of the associated factors (such as the ITV programming change) be evaluated.

The basic step therefore has to be to compare the observed viewing patterns for the news with the predictions of the duplication of viewing law as an interpretative norm. In 1967 the coefficients for this law, for the general run of programmes on different week-days on the two channels, were 1·4 for within-channel duplication on both ITV and BBC1 and 1·0 for between-channel duplication (see Table 4.2).

The news on different days

Table 4.11 shows the extent to which viewers of one news bulletin watched others on another week-day, either two bulletins on the same channel (e.g. ITV × ITV or BBC × BBC) or on two different channels (ITV × BBC).

The results here are again set out in terms of the percentage of the total population viewing both programmes (rather than in the format used earlier, showing the percentage of the audience of one programme who also watched the other). Thus in May, prior to the programming change, on average 9% of the housewife population watched both the 6 pm ITV bulletins on two different days (the first line of figures). Similarly, 9% of the population watched a 6 pm ITV bulletin on one day and a 9 pm ITV bulletin on the other. (This is just over half the audience of the average 6 pm bulletin with a rating of 17, but only a quarter of the larger audience at 9 pm with a rating of 36. Expressing the individual audience duplications in such *relative* terms is often more informative, but it is less succinct when analysing extensive data as here.)

Table 4.11

Observed duplications (O), predictions (P) and deviations (O−P) for news bulletins on different week-days
(averages across the five week-days − 1 to 5 May 1967)

Housewives London & the North	Observed Ratings	% of Population Viewing Both Bulletins		
		O	P	(O-P)
ITV x ITV				
6 pm x 6 pm	17 & 17	9	4	5
6 pm x 9 pm	17 & 36	9	9	0
9 pm x 9 pm	36 & 36	18	18	0
BBC x BBC				
6 pm x 6 pm	15 & 15	6	3	3
6 pm x 9 pm	15 & 24	5	5	0
9 pm x 9 pm	24 & 24	8	8	0
ITV x BBC				
6 pm x 6 pm	17 & 15	3	3	0
6 pm x 9 pm	17 & 24	3	4	-1
9 pm x 6 pm	36 & 15	5	5	0
9 pm x 9 pm	36 & 24	8	9	-1

Comparison with the predicted duplications in Table 4.11 gives clear-cut results. Duplication of viewing for news bulletins on different days is mostly the same as for programmes generally, i.e. it is as predicted by the duplication law.

There are only two apparent exceptions — the high duplications for two 6 pm ITV or two 6 pm BBC bulletins. But even these deviations are predictable. They are part of the high duplications in the afternoon and early evening which occur generally, as described earlier in this chapter. The evidence is that if *other* programmes were screened at 6 pm on two days, they would similarly tend to draw an apparently abnormally high number of common viewers on different evenings.

News bulletins on the same day

Next we turn to the 6 and 9 pm bulletins *on the same day*. These times are far enough apart for the duplication law also to hold without any special "inheritance effect", such as occurs for more or less *adjacent* programmes on the same day (Table 4.8). Nonetheless, the similarity of content of the different bulletins might be expected to lead to abnormal effects.

Table 4.12 shows that this is not so: the same-day viewing patterns were in fact virtually normal. Thus with two ITV programmes rated 19 and 36 and a duplication coefficient of 1·4, one would expect 10% of the population to watch *both* (1·4 X 19 X 36/100 = 10%). That is what occurred.

Table 4.12

Same-day duplications for news bulletins

Housewives London & the North May 1967		Observed Ratings	% of Population Viewing Both Bulletins		
			O	P	(O-P)
6 pm	9 pm				
ITV	ITV	19 & 36	10	10	0
ITV	BBC	19 & 25	5	5	0
BBC	ITV	15 & 36	4	5	-1
BBC	BBC	15 & 25	5	5	1

It followed from these findings that there was no inhibition about watching more than one news bulletin. In fact, the above results imply that of people who watched a 6 pm bulletin (ITV or BBC), about 75% would also watch one at 9 pm. This result is neither specially "high" or "low" — it is just about what would be found for any programmes in these time-slots with the corresponding rating levels. Similarly, about 40% of the audience of a 9 pm bulletin would already have seen a 6 pm one. (The ratings at 6 pm are lower than at 9 pm, as shown in Table 4.12, and hence the duplicated viewers are a lower proportion of the 9 pm audience than of that at 6 pm.)

"News at Ten"

These various results implied that putting on a half-hour ITV news bulletin at 10 pm instead of a shorter one at about 9 pm need lead to no peculiar results. And so it turned out.

Table 4.13 shows that in October 1967, for instance, the percentage of viewers of "News at Ten" among viewers of other news bulletins was close to the predicted levels. A typical "News at Ten" was watched by about the predicted 46% of audiences of other ITV bulletins, and by just under the predicted 31% of the audience of BBC bulletins.

Table 4.13

Duplication of "News at Ten" with other news:
observed (O) and predicted (P)

Adults in London October 1967	% of other News Audience Who Viewed "News at Ten"	
	O	P
ITV News at 6 pm, Same Day	45	46
" " " " Diff. "	43	46
" " " 10 pm, " "	47	46
BBC News at 6 pm, Same Day	28	31
" " " " Diff. "	28	31
" " " 9 pm, Same "	27	31
" " " " Diff. "	30	31

A special result concerns the audience duplication between the BBC 9 pm News and the ITV "News at Ten" on the same day. Previously, when both channels had screened news bulletins at the same time at 9 pm or within a few minutes of each other, watching both was either impossible or required very deliberate switching (which could not in effect be measured by recording of viewing in quarter-hour time-periods). But as the table shows, after the scheduling change, about 27% of the audience of the 9 pm BBC bulletin also watched ITV's "News at Ten" — almost the same as if these were *any* kinds of programmes.

Complementary programming

ITV's change to "News at Ten" occurred with virtually no effect on rating levels as such. But one consequence of the *duplication* results was that slightly fewer people saw at least one news bulletin per day. Previously, with "competitive" programming, somebody watching at 9 pm had to see the news on one channel or the other (few people had BBC2 in 1967). Now, with "complementary" programming on BBC1 and ITV at both 9 and 10 pm, people need not watch the news at *either* time, since they were provided with the choice of a non-news programme on the other channel.

The total number seeing any news therefore drops. This is balanced, in arithmetical terms, by the increased number of people who see *both* bulletins (and indeed, a *longer* one at 10 pm). But the drop in the total numbers seeing news does not imply any special tendency to avoid the news. Similarly, the fact that some people watch both the 9 pm BBC and the 10 pm ITV news does not imply that they are specially avid followers of the news. Virtually the same duplication percentages would have occurred for any programmes with the same rating levels.

Summary

The duplication of viewing law holds for different demographic groups and regions of the country. It also holds for different points in time, although since about 1968 the available viewing data reflect a somewhat greater degree of channel loyalty.

Duplication of viewing between different week-day afternoons or early evenings is relatively higher. This appears to be due to the consistent *non-availability* of those people who are at work, rather than due to any specially intensive viewing by those who actually view. The same high duplication pattern occurs consistently between late-evening programmes on different days, due to fairly consistent bedtime habits by part of the population.

Consecutive or near-consecutive programmes on the same evening share their audience to an above-normal extent, but this "audience inheritance" does not extend to programmes further apart.

There are no special duplication patterns for programmes of any particular type (e.g. comedies, current affairs, etc.). As far as people's actual viewing behaviour is concerned, different programmes of the same type do not appeal specially to the same viewers.

54

5 Repeat-Viewing

We now turn to the subject of repeat-viewing, i.e. the extent of audience overlap for different episodes of the *same* programme, usually screened one week apart. The basic questions are what proportion of the audience of one episode will also watch the next, and on what factors the size of this audience overlap depends, e.g. the type of programme, its popularity or rating, the time of day, etc.

The actual results are simple, but differ in form from those for audience duplication between *different* programmes. We start with an illustration for a series where each episode was at one time screened in two halves.

"Z-Cars"

The popular weekly police programme "Z-Cars" is typical of many TV series: episodes with individual stories but involving the same main characters are shown at the same time each week, here on BBC1 on Monday night from 7 to 7.45 pm. Sometimes however (e.g. in 1967 and again in 1975) a slightly longer script was used and screened in two separate halves each week, on Mondays and Tuesdays from 7 to 7.30 pm.

In a typical week, 24% of London housewives watched the first half of the programme and 24% also watched the second half, as shown in Table 5.1. One might think that these would be the same people, but that was certainly not the case.

Table 5.1

The audience for the two halves of "Z-Cars"

London Housewives 1967	% Viewing
1st half, Monday May 1st	24
2nd half, Tuesday May 2nd	24

The duplication coefficient for BBC1 programmes in general at that time was about 1·4. For two programmes with ratings of 24, the predicted audience duplication would be about $1·4 \times 24 \times 24/100 = 8$; i.e. a *third* of the audience should be in common. But such a prediction of course applies only to two *different* programmes, not two episodes of the same series, let alone two halves of the same episode as here. One might well expect that for a programme like "Z-Cars", the pull of watching the concluding half of a gripping episode the next day would be much higher.

In actual fact, about 12% of the population analysed here (again London housewives) watched both halves of "Z-Cars", as set out in Table 5.2. This is indeed higher than the 8% overlap which the duplication law would predict for different programmes.

Table 5.2

Viewers of BOTH halves of "Z-Cars"

London Housewives 1967	% Viewing
Both Monday and Tuesday 7 - 7.30 pm	12

But in absolute terms, this level of overlap still is low. Only 12 out of 24, or 50%, of the housewives who watched the first half of the programme on Monday also watched — or should one say "bothered to watch" — the second half on Tuesday. And correspondingly, of those who watched on Tuesday, as many as 12/24 or 50% did so without having seen the beginning of the episode the previous day. Yet these are typical findings for repeat-viewing generally, as noted in Chapter 2.

55% repeat-viewing

Table 5.3 gives a more general illustration, namely week by week repeat-viewing results for 40 regularly screened programmes (including film slots and the like) in the spring of 1971, arranged in decreasing order of their rating levels. The striking finding is that generally the percentage of viewers of one episode who watched the next one in the following week is roughly 55%. This holds almost equally for the more popular programmes on the left-hand and for the programmes with lower ratings shown on the right-hand.

Table 5.3

The percentage of the audience viewing again the following week (London housewives, April/May 1971)

	Average Rating	Repeat %		Average Rating	Repeat %
Doctor at Large	35	56	This Week	25	45
Coronation Street	34	63	Bev. Hill Billies	24	62
Sunday Film	34	50	Sportsnight	24	53
This is Your Life	34	68	Cinema	23	49
Dick Emery	32	59	The Virginian	22	60
Ironside	31	58	Val Doonican	22	46
Nearest & Dearest	30	58	Top of the Pops	21	56
Thursday Film	30	58	Golden Shot	21	51
Tuesday Film	30	58	FBI	20	49
Smith Family	30	58	Stars on Sunday	20	49
Two Ronnies	29	50	Top of the Form	20	53
The Western	29	53	Thursday Play	20	42
Opportunity Knocks	29	66	Seven Men	19	49
Budgie	28	65	Please Sir	18	44
Hawaii 5-0	28	65	The Doctors	18	61
The Saint	28	64	Coppers End	17	39
Hine	27	55	Name of the Game	17	50
Persuasion	26	44	The Avengers	15	50
Saturday Film	26	44	Braden's Week	16	48
Match of the Day	25	42	Peyton Place	10	51
AVERAGE RATING: 25			**AVERAGE REPEAT %: 54**		

There is some variation in the individual repeat-viewing percentages but it is not large, mostly from about 45 to 65%, with a few exceptional values. To a first order of approximation, for any of the programmes something like *half* the audience watches the next episode a week later.

This is a very simple finding. It is also perhaps a somewhat remarkable one, in two separate ways. Firstly, that repeat-viewing for this supposedly "compulsive" medium is no higher; secondly, that there appears to be so little dramatic variation between programmes.

It is therefore not common for one programme to attract a vastly more loyal following than another. As we said, there is some variation in the figures in Table 5.3, but apart from a trend with rating levels, there appear to be few other systematic factors at work, as we now show.

Programme types Table 5.4 illustrates that there is little if any systematic difference in repeat-viewing levels for different types of programmes. Most of the results average at about 55%.

Plays have the lowest repeat-viewing level (an average of 42%) in this particular analysis, but it does not seem to be a general result. In other years the repeat-viewing level for different plays in a regular drama slot (e.g. "The Monday Play") has also been over 50% (cf. Table 5.7).

Table 5.4

Average repeat-viewing for different programme types

May 1971	London Men	London H/Ws	Lancs H/Ws	Average
Serials	51	59	62	57
Series	53	57	58	56
Comedy Series	48	51	58	53
Shows	48	54	55	52
Quizzes & Games	58	57	55	57
Sport	47	53	43	48
Films	55	56	52	54
Plays	35	46	44	42
Miscellaneous	52	52	53	52
AVERAGE	51	55	55	54

Demographic factors Repeat-viewing levels have not been found to vary greatly for different demographic groups such as men or women, or different regions of the country, as is illustrated in Tables 5.4 and 5.5.

Week-end viewing There is some suggestion in the analyses so far that repeat-viewing levels for regular week-end programmes are fractionally lower than for week-day programmes, as is illustrated in Table 5.5.

Table 5.5

Repeat-viewing week-ends and week-days

May 1971	London Men	London H/Ws	Lancs H/Ws	Average
Week-end	47	44	52	49
Week-day	55	59	58	57

The difference is not large, but a possible explanation may be in terms of *availability*. Perhaps people are somewhat less settled in their *general* social habits at week-ends. However, more empirical checks are needed here.

Rating levels The only other positive factor found so far is a general tendency for repeat-viewing to decrease with rating levels. For the forty programmes in Table 5.3, this can be summarised in terms of averages for the five programmes with the highest ratings and the five with the lowest:

	Average rating	Average repeat
The 5 highest-rating programmes	34	59
The 5 lowest-rating programmes	15	48

For programmes differing on average by almost 20 rating points, the repeat-viewing percentage therefore drops by just over 10 points. In subsequent work the drop in repeat-level found has, if anything, been bigger.

This is an example of McPhee's "Law of double jeopardy": the fewer people choose an item (a low-rating programme), the less those who choose it "like" it (i.e. low repeat). There are many situations where this law applies (e.g. McPhee, 1963; Shuchman, 1968; Ehrenberg, 1972). (But contrary factors could also be at work, for example that programmes with small audiences could be watched by people who are particularly loyal to them — being perhaps "selective" viewers).

For much higher rating levels a relationship with repeat-viewing level must necessarily occur. Thus two programmes watched by 100% of the population must have 100% of their viewers in common. But just what the detailed quantitative pattern is for audiences at more typical rating levels is not yet altogether clear. More analysis is needed. But it is clear that the relationship is not that of the duplication of viewing law since repeat-viewing level is not directly proportional to rating level. (Hence the effects of rating level may appear less here as we are dealing with variations around 55%.)

Non-stationarity

The analysis of repeat-viewing is basically simple when the rating levels of the two episodes are the same. The two halves of "Z-Cars" analysed earlier provide an example. The rating of each half-episode was 24, and since 12% of the population watched both halves, $12/24 = 50\%$ of the audience of either half also watched the other.

But audience sizes for different episodes of a programme need not be the same. This can obviously happen when one is considering a rather broad type of programme such as a drama or film slot, where the popularity of individual plays or films can vary markedly. But even with a series or a serial there can be variations in audience levels.

A relatively complex illustration is for the twice-weekly ITV serial "Coronation Street" in April 1971. Monday and Wednesday ratings of "Coronation Street" over two weeks among Lancashire housewives were:

Week 1		Week 2	
Mon.	Wed.	Mon.	Wed.
34	39	40	28

The first Monday was Easter Monday, a public holiday with unusual competitive programming and the "Coronation Street" rating was lower than on the following Wednesday. But in the next week the Monday rating was substantially *higher* than that on Wednesday, a pattern which also held up in the following weeks and was no doubt due to popular Wednesday programmes on BBC1.

It follows that whereas in week 1, all the Monday viewers (100%) *could* have watched the Wednesday episode as well, in week 2, *at most* 70% (28/40) of the Monday audience could have done so. These are upper limits to the possible level of repeat-viewing, but they must generally affect the repeat-viewing levels actually attained. One would expect Monday—Wednesday repeat-viewing to be lower in week 2 than in week 1 (and it was about 40% compared with 60%, or the even higher 63% *week-by-week* repeat-viewing for "Coronation Street" then − cf. Table 5.3).

At present it seems that the most stable measure of repeat-viewing would be obtained by using the *smaller* of the two audiences as the base for the percentage. But in this area much more study is needed. (A good deal of further insight into audience flow can probably still be gained, e.g. by the extent to which a change in audience size is due to changes in the number of repeat-viewers of the programme or of viewers of only one episode.)

One practical problem is that variability in audience size (which can at least in part be due simply to sampling errors in the ratings data) has a relatively greater effect on apparent repeat-viewing levels for low-rating programmes than for high-rating ones. Thus for two episodes with high ratings, say 31 and 28, the difference of 3 points can in itself have little effect on the repeat-viewing pattern − as many as 90% (28/31) of those watching the first episode could have watched the second. But for two low-rating episodes with ratings of about 6 and 3, the same difference of 3 rating points will tend to depress the repeat-viewing level — the maximum possible repeat-level based on the first episode is 50%. This could be one factor in the apparent decrease in repeat-viewing with rating level noted earlier.

Irregular programming

So far in this chapter we have discussed the audience overlap between different episodes for the same programme, usually screened at the same time in successive weeks. However, sometimes programmes which are radically different are shown at such times.

It might be thought that when two *different* programmes are shown at 9 pm on successive Mondays say, the duplication of viewing law should apply, just as for the different programmes which are shown on Monday at 9 pm in one week and on Tuesday at 9 pm in the next. But this is not in fact so.

Instead, the same kind of repeat-viewing pattern as for different episodes of the same programme appears to operate. If the two rating levels are the same, generally about half of the audience one week watch the other programme in the following week.

The observed audience overlap between different programmes screened at the same time a week apart is therefore generally higher than that

Table 5.6

The observed audience overlap between various pairs of programmes shown a week apart, and predicted levels from the duplication law

Ratings of Programmes		% of 1st prog. audience viewing 2nd programme	1.7 x rating of 2nd programme
1st	2nd		
29	26	56	44
19	21	48	36
16	17	40	29
15	15	48	26
7	11*	47	19
4	2*	27	3

* Substantial proportional changes

indicated by the duplication of viewing law. The difference is dramatic for programmes with low ratings. Table 5.6 gives an illustration for some typical cases in 1969 where the two rating levels were mostly more or less similar.

It therefore appears that repeat-viewing in successive weeks is influenced more by general social habits and by "availability" — e.g. whether one is actually watching TV at all — than by the particular programmes shown. This is at least so in the context of fairly regular programme scheduling (especially on ITV).

Viewing the other channel

If only about 55% of the audience of a programme usually watch the next episode of the programme, what are the remainder of the earlier audience then doing? Do they still watch television but on another channel (possibly because of the conflicting preferences of other family members)? Or do they not watch television at all (and is this more or less unavoidable, like being out, or only due to not feeling like it)?

The answer so far available is that only a small percentage of the "lapsed viewers" watch television on another channel at that time (at the

Table 5.7

Viewing the same or the other channel a week later
(ITV and BBC1)

London Housewives June 1969	% of audience who exactly a week later watch		
	SAME Channel	OTHER Channel	EITHER Channel
Programme-type			
Twice-weekly Serials	62	8	70
Other Serials	55	5	60
Series	55	12	67
News at Ten	55	11	66
Documentaries	48	14	62
Quizzes and Games	57	6	63
Variety Shows	52	13	65
Magazine Programmes	51	11	62
Plays	51	9	60
Films	52	16	68
Low-Rating Programmes	32	12	44
AVERAGE	52	10	62

time, the third channel, BBC2, was received and viewed only to a very small extent). Table 5.7 illustrates for data in June 1969 that whilst 50% tend to be repeat-viewers, of the rest only about 10% watch the other channel instead. Thus, the main factor in *non-repeat-viewing* is not watching television at all.

Non-consecutive episodes

The preceding findings — i.e. that most non-repeat-viewers have not changed to another channel and that there is mostly relatively little variation in the repeat-viewing loyalty of different programmes — should imply that those viewers who do not watch the next episode of a programme are not in fact "lost for good". It is less a reaction against the particular programme than a reflection of irregular viewing habits. They may well watch the programme again the week after that.

This is borne out by the facts. There is little if any erosion in the percentage of repeat-viewers for episodes further than a week apart. Table 5.8 illustrates this for all programmes (about 35 on ITV and BBC) screened regularly (at the same time each week) over the four-week period in April/May 1971 already analysed.

Table 5.8

Repeat-viewing in four successive weeks for regular programmes
(regular programmes, April/May 1971)

% of 1st Week audience who viewed in :	London Men	London H/Ws	Lancs H/Ws	Average
second week	50	55	56	54
third week	49	50	55	51
fourth week	49	52	52	51
AVERAGE	50	52	54	52

The average figures stay at over 50%. The slight decline in weeks 3 and 4 is not a general feature, i.e. it is not borne out by analyses of data at other times. Instead, it is due to a slight seasonal decline in rating levels from week 1 onwards at that time.

"Mr Trimble"

The implication of the repeat-viewing results just discussed is that those people who watch a particular programme do not generally watch every episode of it.

As an example we consider an ITV programme of twenty-five episodes, "Mr Trimble", for children of preschool age. This was screened in "strip-programming", i.e. every week-day between 12.05 pm and 12.25 pm from 5 February to 9 March 1973.

Direct measurement of preschool children's viewing is not routinely available, but an assessment of their *maximum* exposure to the series could be made by analysing whether the TV set was switched on to ITV in homes with at least one child aged five or under. In these homes, the rating for "Mr Trimble" was about 20, so that 1 in 5 saw the typical episode.

However, during the five weeks just over 60% of homes saw at least one episode (and on average, these 60% saw about eight episodes each, or one in three). Repeat-viewing for any *two* episodes was at the normal level for programmes generally, namely 53% (but was fractionally higher, about 60%, for episodes which were consecutive or at least in the same week). The detailed frequency distribution of exposure is set out in Table 5.9.

Table 5.9

Number of episodes of "Mr Trimble" seen
(in homes with children under 5)

1973	0	Number of Episodes Seen																								
		1	2	3	4	5	6	7	8	9	10	11	12	13	14	15	16	17	18	19	20	21	22	23	24	25
% of homes	38	17	6	2	3	2	3	3	2	2	3	1	3	2	2	2	1	1	1	0	2	1	1	1	0	1
(Grouped)	38	—— 33 ——						—— 14 ——						—— 9 ——						—— 6 ——					

These results show that for an extensive programme series with a typical repeat-viewing level (about 55%), most of its viewers saw only a few episodes, only 1 in 10 saw three out of four, and almost none saw all.

Audience cumulation

Although few programme series as extensive as "Mr Trimble" have as yet been analysed, the results appear more or less typical. This is shown by analyses of once-or-twice-weekly programme series screened over four weeks (i.e. four or eight episodes in all).

In general, relatively few viewers see all or nearly all the episodes in a series, although in analysing any *shorter* series of episodes, the effect is not quite as dramatic as for the twenty-five episodes of "Mr Trimble". (For the five episodes of "Mr Trimble" shown *in a single week,* as many as 25% of those seeing the programme at all that week saw either four or all five of the episodes).

We now describe typical results for once and twice-weekly programmes. These follow regular patterns which can be summarised by a theoretical model, the "BBD".

As a start, we consider four successive episodes of "This is Your Life" in the four weeks ending 9 May 1971. The ratings amongst London men were about 32, with small irregular variations of about 1 percentage point. On average, 69% of the viewers of one programme also saw another (whether in an adjacent week or not). This is a relatively high repeat level compared with most programmes.

Table 5.10 shows that 50% of men saw *none* of the four episodes, 15% saw *one* of the episodes, 9% saw *two,* 13% saw *three,* and another 13% saw all *four.*

Table 5.10

The distribution of exposures to "This is Your Life"

London Men April/May 1971	% Seeing the Programme				
	0	Once	2 times	3 times	4 times
All Men = 100%	50	15	9	13	13

(The rating for the average or typical episode is made up of a quarter of those seeing only one episode, i.e. 15/4, plus half those seeing two out of 4, i.e. 9/2, etc., giving $4 \cdot 7 + 4 \cdot 5 + 9 \cdot 8 + 13 \cdot 0 = 32$.)

Reach and average frequency

Two conventional concepts in dealing with such distributions of the frequency with which such a "schedule" of television broadcasts is seen are *reach* and *average frequency.*

Reach is the percentage (or number) of the population who are reached at all by the schedule, i.e. those who see at least *one* of the broadcasts in question. The reach of "This is Your Life" over four episodes was therefore 50%. Thus although about 32% of men saw any particular broadcast, viewers of different episodes were not always the same people and hence a total of 50% saw *at least* one episode.

Average frequency is the average number of episodes seen by those people reached at all. If people watching one episode also saw all the other three, the average frequency would clearly be 4. In our example, the total number of episodes seen is built up from the 15% who saw one, the 9% who saw two, and so on, giving

$$15 \times 1 + 9 \times 2 + 13 \times 3 + 13 \times 4$$
$$= 124 \text{ episodes}$$

seen per 100 London men. But since only 50% of London men saw *any* of the four broadcasts, the average frequency with which they did so was

$$124/50$$
$$= 2 \cdot 5$$

Thus half the population saw "This is Your Life" in the four weeks (the "reach"), and they each saw on average 2·5 out of the four episodes.

The higher the repeat-viewing percentage, the lower is the reach and the higher the average frequency of exposure per viewer (i.e. the more the *same* people view each time). A programme with an average rating of 20 and a typical repeat-viewing percentage of 55 would reach as many as 60% of the population in four episodes, with each viewer seeing an average of' two episodes.

Theoretical models

Dealing with a variety of such frequency distributions becomes a relatively complicated matter because of the number of variables included, such as the number of episodes, the rating levels and the repeat-viewing levels. A mathematical model to summarise such data can therefore be helpful, e.g. for prediction.

This type of knowledge will generally be of interest to those responsible for programme decisions and to students of mass communications. A more specific motivation for such research is that it provides advertisers and agencies with important information for media planning. Given an advertising campaign of so many "spots", the advertising agency and its client would usually like information not merely on total ratings (obtained from the regular syndicated research) but also the number of people exposed to 0, 1, 2, etc. of the advertisements. If a predictive model can be developed not only can past campaigns be evaluated, but alternative future plans can be compared.

There have been a number of efforts to produce suitable formulae. In the UK one of the earliest was that produced by JICTAR's previous research agency, Television Audience Measurement. In 1966 TAM produced a reach and frequency guide based on 280 schedules which were evaluated in 1964 and 1965. The guide permitted reach and average frequency to be predicted from information on the average rating of the schedule and the maximum rating.

By 1969 it was apparent that the observed reach and frequency in most schedules differed from that predicted by the 1966 TAM guide. This stimulated further work by various parties to improve the TAM predictors. In 1972, Audits of Great Britain (AGB), which had succeeded TAM in holding the JICTAR research contract, produced its own guide (Fawley and Fairclough, 1972). Other models were developed by Johnson and Peate (1966), Barnett and Lougher (1971), and Hulks and Thomas (1973). The last uses the Negative Binomial Distribution (NBD) to explain the distribution of frequencies.

A somewhat more direct approach is available for explaining the reach and frequency associated with regularly screened programmes. This approach estimates the frequency distribution of viewing from a Beta-Binomial Distribution, or BBD for short. This model is based on two notions:

1 That the probability that a given person will watch that week's episode of the programme is constant from week to week (for that person) and independent of whether or not he has watched it in previous weeks. (This is supported by the sort of results illustrated in Table 5.8.)
2 That the numerical value of this probability varies from person to person (as shown by the fact that some viewers watch most episodes, and some only a few) and that this variation follows a so-called Beta-distribution in the whole population of potential viewers.

The Beta Distribution has no direct justification at this stage, but it is a "flexible" distribution which can take a variety of different shapes and hence is not a very restrictive assumption. (An earlier assumption of a similar kind in the field of consumer purchasing behaviour has recently been fully justified — see Goodhardt and Chatfield, 1973.)

The consequence of these assumptions is a mathematical formula, the BBD (described in the mathematical appendix at the end of this chapter). With it we can calculate, from the average rating and the average repeat-percentage, how many people saw 0, 1, 2, 3, etc. episodes, and in particular the reach and average frequency of viewing.

The use of a BBD for media exposures was first introduced by Hyett (1958) and extended by Metheringham (1964) for *print* media, and has been taken up by a number of US workers. It is applied here to TV programme schedules under "stationary" conditions (i.e. when there is little or no variation in rating levels or repeat-levels) and also extended to frequency distributions for non-stationary situations and "mixed" schedules.

The fit of the BBD model

For exposure distributions to different episodes of a programme under stationary conditions — when rating and repeat-viewing levels show little or no variation — the BBD model generally gives a good fit.

This is illustrated in Table 5.11, both for the schedule of "This is Your Life" in the four weeks ending 9 May 1971 (shown in Table 5.10) and for the average of some twenty other "stationary" four-week schedules (covering serials, series, comedy and musical shows, quizzes, films, plays, etc.).

Table 5.11

The fit of the BBD to stationary schedules of four episodes
(observed and theoretical BBD figures)

London Men April/May 1971		% Seeing the Programmes				
		0	Once	2 times	3 times	4 times
This is Your Life	Obs.	50	15	9	13	13
	BBD	49	15	11	10	15
Average of 20 cases	Obs.	57	16	10	9	8
	BBD	57	16	10	9	8

The model therefore serves to summarise such data succinctly. Given the average rating and repeat-viewing level, it can reproduce the full frequency distribution of exposures and, in particular, the reach and average frequency. Thus for "This is Your Life", the observed and predicted reach are 50 and 51, and the average frequency values are 2·48 and 2·49.

Non-stationary schedules

The BBD model also serves to summarise many *non-stationary* cases to a first order of approximation, even though the theoretical basis from which the model was derived no longer applies.

Table 5.12 summarises results for more than a dozen regularly screened programmes whose ratings declined over the four weeks analysed by an average of 25%. The observed frequency distributions of exposures are still adequately summarised by the theoretical BBD (to within an average of 1 percentage point for the various individual schedules).

Table 5.12

The fit of the BBD for some non-stationary
schedules of four episodes

London Men April/May 1971		% Seeing the Programme				
		0	Once	2 times	3 times	4 times
Average of 14 non-stationary programmes	Obs.	58	18	11	8	5
	BBD	57	19	11	8	5

The value of such analyses of the distribution of exposures to a series of episodes is that it allows us to summarise, and hence understand, such viewing patterns better. For the typical programme in Table 5.12, viewed by just over 20% of the population in an "average" week, we can draw conclusions such as the following:

(i) that twice as many people (i.e. $100 - 58 = 42\%$) would see it at least once in four weeks;
(ii) that only a quarter of a given episode's audience — 5% out of 20% — sees all four episodes (or only about $5/42 = 12\%$ of those reached in the four weeks); and
(iii) *that such results are normal and predictable for the general run of programmes.*

Furthermore, a descriptive model like the BBD, which works under a wide range of conditions, has value even in those particular cases where it does *not* fit the data, by pointing to the nature of the discrepancies. An illustration of this is provided by the analysis of the twice-weekly ITV programme "Coronation Street" in Table 5.13.

69

Table 5.13

The observed and theoretical distribution for eight episodes of "Coronation Street" in four weeks

London Men April/May 1971		Number of Episodes Seen								
		0	1	2	3	4	5	6	7	8
Coronation Street	Obs. %	41	13	8	7	8	6	6	6	5
	BBD %	37	15	10	8	7	6	6	5	5
Difference		4	-2	-2	-1	1	0	0	1	0

The average rating was 29, and the average repeat-viewing percentage 59. The theoretical BBD for such a schedule gives broadly the right picture, but it differs significantly in some details. Somewhat fewer people were reached than predicted (59% versus 63%), because fewer saw only one or two episodes (21% versus 25%). These discrepancies are probably due to the *non-stationarity* for this series of episodes noted earlier (Easter Monday, etc).

Mixed schedules

The BBD model can also describe audience cumulation for even more "mixed" schedules. Instead of series of time-slots (or programmes) at the same time of a certain day each week, as has mainly been considered so far, a mixed schedule might consist of Monday at 8.30, Wednesday at 10, Thursday at 8.45 the next week, and so on. Such mixed schedules can arise in considering *types* of programmes (e.g. exposures to different plays, or to the news), and more particularly also in considering advertising campaigns.

Table 5.14

The frequency distribution of viewing seven different programmes (one programme between 7 and 10 pm for each day of the week)

London Men 1971		Numbers of Programmes Seen							
		0	1	2	3	4	5	6	7
Mixed Schedule	Obs. %	27	23	20	12	9	5	3	1
	BBD %	28	23	18	13	9	6	3	1

Table 5.14 gives a fairly typical example of a mixed schedule for seven ITV programmes, one per day throughout a week. The empirical result is once more that few people see all or nearly all the programmes. As long as we exclude cases of two programmes shown close together on the same day, the BBD model gives a good fit again.

Summary

Only about half the people who see a repetitive programme one week see the next episode in the following week. There is little difference in this by type of programme or demographic group. The repeat-viewing level shows little change from the average level of about 55% even for episodes further than one week apart. But there is some tendency for the repeat-viewing level to increase with the size of the audience.

It follows that few people see all or nearly all the episodes in any extended series.

The implication is that failure to repeat-view is generally a matter of variable social habits rather than a reaction to programme content.

MATHEMATICAL APPENDIX: THE BETA-BINOMIAL MODEL

The use of a Beta-Binomial Distribution (BBD) to describe the proportion of the potential audience who see $0, 1, 2, 3, \ldots, n$ episodes out of a series of n consecutive showings of a regularly screened programme derives from the two assumptions noted on page 67, namely that

1   the probability that a given person will watch a particular week's episode of the programme is constant from week to week (for that person) and independent of whether or not he has watched in other weeks, and

2   the numerical value of this probability varies from person to person and has a Beta Distribution in the whole population of potential viewers.rs.

The first of these assumptions implies that for a person who has a (fixed) probability p of seeing each of the episodes, the probability that he will see exactly r out of the n episodes is given by the Binomial Distribution. Thus:—

$$\text{Prob } (r|p) = \frac{n!}{r!(n-r)!} \, p^r (1-p)^{n-r}, \quad 0 \leqslant r \leqslant n$$

The second assumption states that the probability p is distributed through the population with a probability density function given by

$$f(p) = \frac{p^{\alpha-1}(1-p)^{\beta-1}}{B(\alpha, \beta)}, \quad 0 \leqslant p \leqslant 1,$$

where α and β are positive constants, and $B(\alpha, \beta)$ is the Beta function

$$B(\alpha, \beta) = \int_0^1 x^{\alpha-1}(1-x)^{\beta-1} dx$$

Putting these two assumptions together implies that the probability $P(r)$ that a person chosen at random from the population will see exactly r out of the n episodes is given by:

$$P(r) = \int_0^1 \text{Prob } (r|p) f(p) \, dp$$

$$= \frac{n!}{r!(n-r)!} \frac{1}{B(\alpha, \beta)} \int_0^1 p^{r+\alpha-1} (1-p)^{n-r+\beta-1} dp$$

$$= \frac{n!}{r!(n-r)!} \frac{B(\alpha+r, \beta+n-r)}{B(\alpha, \beta)}$$

The distribution defined by this equation is called the Beta-Binomial Distribution. It follows from elementary probability theory that in a large population the proportion of the total population who see exactly r out of n episodes will also be $P(r)$.

The BBD is defined by three parameters: the number of episodes, n, and the two constants α and β. In any practical application, therefore, the two parameters α and β have to be estimated. It can be shown that under the two assumptions of the BBD model given above, the proportion of the population seeing each episode (i.e. the rating of each) is equal to $\frac{\alpha}{\alpha+\beta}$, and the proportion of the audience of one episode who see a particular other episode (i.e. the repeat-viewing rate) is equal to $\frac{\alpha+1}{\alpha+\beta+1}$ Thus the parameters can be estimated by solving the two equations:

$$\text{average rating} = \frac{\alpha}{\alpha+\beta}$$

$$\text{average repeat} = \frac{\alpha+1}{\alpha+\beta+1}$$

Substituting these estimated values of α and β into the equation for $P(r)$ for different values of r gives the complete frequency distribution for the n episodes. In particular with $r = 0$, $P(0)$ is the proportion seeing no episodes and so the reach is calculated as $1 - P(0)$.

The two parameters α and β of the BBD make it a very flexible distribution over the integers from 0 to n. For various values of the parameters, it can take a wide variety of shapes from ordinary humped shapes, either symmetrical or skew, to J and reverse-J shapes and even U shapes.

Mixed distributions

Because of the great flexibility of the distribution, it has also been found to provide (in most circumstances) a good fit to the distribution of the number of programmes seen out of a mixture of n different programmes. In this case the first assumption referred to above clearly does not apply (since the different programmes may have radically different ratings) and so the second assumption has no meaning. There is therefore no theoretical model explaining why the BBD fits; it is just a mathematical convenience.

In this case the estimation of the parameters α and β in practical cases is rather more complex. It depends on the fact that the mean of the distribution of the number of programmes seen must equal the sum of the rating of each of the programmes, and the variance of the distribution can be calculated from the sum of the duplications of all pairs of programmes. The mean and variance so calculated are then equated to the mean μ and variance σ^2 of the BBD given by:

$$\mu = \frac{n\alpha}{\alpha + \beta}$$

$$\sigma^2 = \frac{n\alpha\beta(n + \alpha + \beta)}{(\alpha + \beta)^2(1 + \alpha + \beta)}.$$

and the two equations solved to give estimates of α and β. The frequency distribution is then calculated according to the equation for $P(r)$.

6 Intensity of Viewing

We have noted that hardly any viewers see all or nearly all episodes of a regular programme shown over a period of time, while most see only a few. More generally, some people are heavy consumers of television and others view relatively little.

In this chapter we examine this variation in people's "intensity of viewing". What differences are there in their amounts of TV viewing, and how does this relate to choice of channel and programmes, and to repeat-viewing and duplication patterns?

Hours viewed

The average number of hours of TV viewed in the UK in a week is generally almost 20 hours for women and slightly less (say 17 hours) for men. The figures tend to be a little lower in the summer than in the winter.

There is however a good deal of variation about these averages. If we divide people into three equal groups according to how much television they watch in a week, then a typical distribution (for London housewives in April 1971) is that the third who were the heaviest viewers watched on average about 30 hours, and the third who were lightest viewers watched about 10 hours on average:

Heaviest third — 30 hours TV on average per week
Medium third — 20 hours TV on average per week
Lightest third — 10 hours TV on average per week

Table 6.1

The number of hours TV viewed per week

London Housewives 1971	Number of Hours Viewed												
	0	1-	4-	7-	10-	13-	16-	19-	22-	25-	28-	31-	34-
% of population	0	1	6	4.	8	13	11	14	15	9	8	6	5

The distribution is shown in more detail in Table 6.1 in three-hourly groupings. In particular, just over 10% of housewives viewed *less* than 10 hours a week; just over 10% viewed more than 30 hours.

Channel choice

Total TV viewing in the UK usually tends to be split roughly 50:50 between ITV and BBC. For example, in April 1971 London men in households with TV sets capable of viewing both ITV and BBC1 (and possibly BBC2) transmissions viewed ITV on average for 9 hours and BBC for 8 hours per week (7 hours BBC1 and 1 hour BBC2). London housewives on average viewed 2 more hours of television, these being on ITV (11 hours ITV, 7 hours BBC1 and 1 hour BBC2).

If we examine this split in terms of light, medium and heavy viewers, heavy viewers of television watch more ITV (the channel with on the whole the more "popular" programmes). Table 6.2 shows that the heaviest TV viewers on average watch about twice as much ITV as BBC1 (19 versus 9 hours), whereas the light viewers watch about equal amounts (4 hours of each). All groups watch BBC2 to about the same (small) amount of 1 hour on average. In broad terms these patterns are general ones, with the partial exception of the results for BBC2 which at the time — 1971 — could be received by a smaller proportion of TV sets than now.

Table 6.2

Channel choice among heavy, medium and light viewers of television
(Average hours viewed per week)

London Housewives 1971	TV Viewers			
	Heavy (33%)	Medium (33%)	Light (33%)	All (100%)
Av. hours viewed	Hours	Hours	Hours	Hours
ITV	19	10	4	11
BBC1	9	8	4	7
BBC2	1	1	1	1
Any TV	29	19	9	19

Cross-tabulating the intensity of viewing on each of the two main channels as in Table 6.3, we see that there is a tendency for heavy viewers of one channel to be fairly light viewers of the other. The tendency is not large, but the *opposite* pattern (heavy ITV viewers mostly being heavy BBC viewers) certainly does not exist. Thus the 33% of housewives in the first column of Table 6.3 who are heavy viewers of BBC1 are made up of only 5 percentage points who are heavy ITV viewers and as many as 14 who are medium and 14 who are light ITV viewers (instead of an even 11, 11 and 11 in each category as would occur if the two types of viewing were unrelated).

Table 6.3

Heavy, medium and light viewers of ITV and BBC1
(% viewing both channels)

London Housewives 1971	ITV Viewers:			Average %
	Heavy (33%)	Medium (33%)	Light (33%)	
BBC1 Viewers:				
Heavy (33%)	5%	14%	14%	11%
Medium (33%)	13%	11%	9%	11%
Light (33%)	15%	8%	10%	11%
Average %	11%	11%	11%	11%

A simple interpretation of these results is that apart from a short fall of people who are heavy viewers, as defined here, of both ITV and BBC1 (there are not enough hours in the day), the viewing intensities on the two main channels are broadly uncorrelated.

Programme choice

A common view of television audiences is that people who do not view much must be selective in what they watch. The word "selective" here would usually be taken to imply that these people watch minority programmes, especially perhaps ones with a cultural or specialist appeal, which therefore only attract small audiences, rather than the popular (and high-rating) type of entertainment programmes.

In practice, this does not occur. If anything, light viewers tend to watch the popular programmes — that is one reason these programmes have high ratings. In contrast, heavy viewers tend to "watch anything", high *and* low-rating programmes. That is why heavy viewers are heavy viewers.

Numerically the differences here cannot be dramatic. Heavy viewers must make up the larger part of the audience of almost any programme. Light viewers must usually be only a small part of each audience. In fact, the heavy viewers, defined here as making up 33% of the population, account for about 55% of the audience for the average TV programme, as shown in the first line of Table 6.4. Similarly, the light viewers' group, again 33% of the population as a whole, only account for some 15% of the viewing audience for the average programme.

Table 6.4

The importance of heavy and light TV-viewers for
high and low rating programmes

London Housewives 1971	TV Viewers			
	Heavy (33%)	Medium (33%)	Light (33%)	All (100%)
The average programme	55	30	15	100
High-rating programmes (35+)	50	30	20	100
Low-rating programmes (-20)	65	25	10	100

Against this background we can now examine the make-up of the audience of high and of low-rating programmes (defining these as having ratings of 35 or more, and of 20 or less respectively). Table 6.4 shows that heavy viewers generally account for 65% of the audience to the low-rating programmes but only 50% of the high-rating ones. The opposite pattern occurs for light viewers. They make up only 10% of the audiences for low-rating programmes but as much as 20% of the high-rating ones.

These results therefore go directly counter to the notion of light viewers being intellectually "selective". In fact they tend, if anything, to select the more *popular* programmes to watch!

Repeat-viewing

Another feature of "selectiveness" might be that light viewers are more *regular* viewers of the programme they watch week by week. They may not watch much, but perhaps they know what they like.

However, the actual results again go in the contrary direction. In Chapter 5 we saw that a little over 50% of the audience of a typical programme would see it again in the following week. But for light viewers, the repeat-viewing percentage is lower, not higher, at about only 40%.

The lower repeat-viewing level is largely related to the level of the ratings themselves. In Chapter 5 we noted a trend whereby the lower-rating programmes also tended to have lower repeat-viewing levels. It looks as though the lower repeat-viewing among light viewers may merely be in line with this (ratings among light viewers are generally low). If the repeat-viewing figures were adjusted for rating levels amongst light viewers, it may therefore be that their repeat-viewing is just like anyone else's, i.e. "normal". But full quantitative details in this area are not yet established. However, it is already quite clear that light viewers are hardly exceptionally *regular* viewers of the programmes they watch.

Duplication of viewing

The duplication of viewing law, which was discussed in Chapters 2 to 4, states that viewers of one programme (X) are more likely to watch another programme (Y) than is the population as a whole, if the two programmes are on the same channel. Thus if 10% of the whole population watch Y, then about 17% of the audience of X would watch Y.

The reason for such above-normal degree of audience duplication for two programmes on the same channel lies essentially with the heavy viewers of that channel – they tend to watch *both* programmes. Thus the audience of programme X will generally contain more heavy viewers than exist in the population as a whole – Table 6.4 showed that of the order of 50 to 65% of the audience would be heavy viewers, compared with the 33% heavy viewers who (by their definition) occur in the population. Correspondingly, the audience of programme Y will tend to contain 50 to 65% heavy viewers. Hence both programmes contain an above-average proportion of heavy viewers and there should be more than average duplication (i.e. a duplication coefficient greater than 1) between the two audiences.

It follows that there should be no above-normal audience duplication within any population subgroup which is more or less homogeneous in its viewing intensities for that channel, i.e. a subgroup where different people all view about the same amount.

Table 6.6 shows that this is what tends to occur. It gives the audience duplication between the six quarter-hour segments starting on the hour from 6 to 11 pm for a Monday and Wednesday *among the 25% heaviest ITV viewers*. (A narrower definition of "heavy viewers" has been used here, so as to obtain a more homogeneous group.) The duplication percentages are high, mostly in the 50's to 70's (since heavy ITV viewers watch many different ITV programmes), but so are the programme ratings themselves among these heavy viewers. Indeed, the two sets of figures for the Wednesday programmes are virtually equal, averaging at 68 and 65 respectively, and so showing *no* tendency towards positive duplication, just as predicted. Thus 58% of the heavy viewers viewed on Wednesday at 6 pm, and about 60% of any of the Monday audiences watched on Wednesday at 6 pm (except for the usual 6 pm X 6 pm blip). And so on for the other times — the discrepancies between the average duplications and the heavy-viewer ratings average at less than a percentage point.

Table 6.5

Duplication of viewing amongst the 25% HEAVIEST ITV viewers

Heavy ITV-Viewing London Housewives 1971	Who also viewed ITV on WEDNESDAY at					
	6 pm	7 pm	8 pm	9 pm	10 pm	11 pm
Viewers of ITV on MONDAY at						
6 pm	(82)*	87	82	74	74	53
7 pm	65	78	76	70	67	42
8 pm	63	78	76	70	67	43
9 pm	59	72	72	74	69	44
10 pm	61	75	80	75	77	55
11 pm	54	69	81	81	81	(69)*
Average Duplication	60	76	78	74	72	47
Rating**	**58**	**75**	**80**	**74**	**70**	**43**

* Early and late evening duplications excluded from the averages
** Amongst these heaviest viewers

This illustrates how there is no above-normal audience duplication among a group of viewers who are relatively homogeneous in their viewing intensities. Viewers of a programme are no more likely than other members of the group to watch another programme. These results are typical of what has been found more generally for heavy and medium-heavy viewers.

Only amongst the *lightest* viewers is there still some substantial positive duplication. Here a relatively large and hence heterogeneous group had to be analysed for statistical sampling reasons, namely the lightest 50%. The number of hours viewed per week still ranges quite widely, from 0 to about 10 ITV hours per viewer. Hence the heavier viewers in this group should still contribute to a positive audience duplication. More analysis on larger samples and more homogeneous subgroups is therefore still needed here to establish the facts for light viewers firmly.

A theoretical model

The results just given serve to confirm an earlier theoretical model of programme choice (Goodhardt, 1966) which was developed soon after the duplication of viewing law was first empirically established.

The model is of a stochastic or "as if random" form. It allows for people's differences in viewing intensity and also supposes that they differ from each other in their individual patterns of programme choice, so much so in fact that their choice patterns look effectively "as if they were random". Under this model the audience at any time t is regarded as generated by sampling the ith individual from the population with a probability which is related first to the audience size (or rating) r_t at time t and second to the individual's general intensity of viewing, v_i say. The latter quantity can be defined as the total number of hours viewed per day by the ith individual, divided by the daily hours viewed by the *average* individual; it does not vary with the programme being shown.

Thus the probability p_{it} of the ith individual viewing the tth time-segment is given as a first approximation by the equation $p_{it} \doteqdot v_i r_t$. (This ignores the fact that this can give some probability estimates numerically greater than one and that the sampling in the model should be "without replacement" from a finite population, but the effect of this is small.)

If for any two times t and s on two different days we now suppose that the sampling of the ith person can be regarded as independent, then r_{st}, the proportion of the population of n individuals who view at *both* times (i.e. the duplicated audience), should be given by

$$r_{st} = \text{Sum}\,(p_{is} p_{it})/n$$

But since $p_{it} \doteqdot v_i' r_t$ and $p_{is} \doteqdot v_i'' r_s$, where v_i' and v_i'' are the ith individual's relative viewing intensities on the two different days, this gives

$$r_{st} = \left\{ \text{Sum}\,(v_i' v_i'')/n \right\} (r_s r_t)$$

The summation term on the right-hand side is the same for all pairs of times s and t on the two days, i.e. it is a constant. This theoretical result therefore agrees with the empirically observed duplication of viewing law $r_{st} = kr_s r_t$, where k is a constant.

The constant k can be calculated either from the observed duplications r_{st} (as was generally done in Chapters 2 to 4) as $k = \text{Sum } r_{st} / \text{Sum } (r_s r_t)$, where the summation is over all relevant times s and t on the two days. Alternatively, k can be calculated by using only the viewing intensities v_i' and v_i'' per day for all the $i = 1$ to n individuals, namely as $k = \text{Sum } (v_i' v_i'')/n$. It can be shown that these two expressions are mathematically identical.

This theoretical argument largely explains the observed duplication of viewing law. When some people view much more than others and this pattern is more or less constant from day to day (or week to week), the sum of the cross-products $\text{Sum } (v_i' v_i'')/n$ will be greater than one, as is found for within-channel duplication of viewing. For a group with *homogeneous* viewing intensities, the values of the v_i — the ratio of the ith individual's viewing hours to those of the average individual — will be approximately one. Hence k also will be about one (as was illustrated in Table 6.5).

Again, for cross-channel duplication, heavy viewers of ITV are *not* heavy viewers of BBC1 (Table 6.3). If anything the opposite occurs. People who are "heavy viewers" on one channel are relatively light viewers of the other. This explains why the cross-channel duplication coefficients are fractionally less than one, as reported in Chapter 3.

Summary

Viewers differ greatly in the amount of television which they tend to watch. Something like 10% watch more than 30 hours a week (or 4 hours a day), and at the other extreme 10% watch less than 10 hours a week (an hour or so on the average day).

Heavy viewers watch more ITV than BBC. They also make up a disproportionately large portion of the audience of low-rating programmes (they have to watch both high and low-rating programmes in order to be heavy viewers).

Light viewers, in contrast, tend if anything to watch the popular, high-rating programmes (which is why they are popular). Light viewers therefore certainly do not generally choose to watch specialist (i.e. low-rating) programmes. They also show no sign of being more regular viewers of the programmes that they watch. They do not appear to be "selective" in either sense, but just to watch less.

Within any subgroup with similar viewing intensities (i.e. average hours of TV viewed), there is little or no positive duplication between audiences on the same channel. The explanation of the duplication of viewing law lies in the variability in the amounts people view, coupled with their variability of programme choice.

7 Audience Flow in the US

In this chapter we describe some comparable results about audience flow in the United States. The analyses are relatively limited but illustrate that the approach of the preceding chapters developed for the UK can also be applied to the study of television audiences elsewhere. In any case, far less work is now needed to establish whether or not the same results recur. One does not have to reinvent the wheel afresh every time.

Despite differences in the US and UK television scenes, the finding is that the main patterns of audience flow checked so far are the same as in the UK. For example, the duplication of viewing law between different programmes operates, both ten years ago in relatively "small" regions such as Birmingham, Alabama and Las Vegas, which then had only two or three main channels (e.g. Ehrenberg, 1966), and in 1974 in the New York City area, where up to 10 or so channels could be received.

Additional results are that repeat-viewing of different episodes within a week is 55% for the three national networks, roughly the UK level, and that the "inheritance effect" operates for consecutive programmes on the same day.

So far there are no analyses of week by week repeat-viewing for the US, since the Arbitron data analysed here cover only one week's viewing for each informant (as mentioned in Chapter 1). Nor have the US results been cross-analysed by heavy and light viewers (as in Chapter 6 for the UK), or week-end viewing been systematically tackled.

We illustrate these findings for the US with the most up-to-date results available, from a sample of 1779 female heads of household in the New York City area in January and early February 1974. (The sampling was spread over four weeks, but only one week's viewing is measured per informant, as already mentioned.)

The results relate to the six most popular stations in the New York City area. These are the affiliates of the three national networks (CBS, NBC and ABC), plus three stations (WOR, WNEW and WPIX) which either belong to regional networks or are independent. Ratings for the educational station WNET were too low to be usefully included in the analysis so far.

Repeat-viewing of different episodes in the same week

We start with an analysis of the repeat-viewing of different episodes of the same programme. The situation analysed here is that of strip-programming, i.e. different episodes of the same programme being shown on successive week-days. This occurs in two different ways. Firstly, repetitive programmes are shown from 4 to 7.30 pm on the three national networks, if we also count local and general news in this category. Secondly, the other three stations show reruns of old programmes and old films in the afternoon and in some cases throughout the evening, as well, e.g. repeats of "The Lucy Show" originally shown on NBC twenty years ago.

On the national networks repeat-viewing for different week-days averages at roughly 55% — the same level as found in the UK for week by week repeat-viewing. Table 7.1 shows the extent to which episodes on a Friday were watched by viewers on other week-days. The overall average is

Table 7.1

Repeat-viewing within the week: NETWORKS
(% of Monday to Thursday audiences of a programme who watched the Friday episode of that programme)

New York Housewives Jan-Feb 74			% Viewing on Friday of the Audience on				
			Mon	Tue	Wed	Thu	Av.
WCBS	4.00	Secret Storm	72	65	69	69	69
	4.30	Mike Douglas	65	64	61	71	65
	6.00	Ch 2 News-6	64	67	60	72	66
	7.00	CBS Eve News	62	63	63	65	63
	Average		66	65	63	69	66
WNBC	4.00	Somerset	71	76	66	68	70
	4.30	Movie Four	15	20	18	29	21
	6.00	Sixth Hour	51	50	51	58	52
	7.00	NBC Night News	45	50	45	49	47
	Average		45	59	45	51	47
WABC	4.00	Love Am Style-D	38	47	39	56	45
	4.30	4.30 Movie	29	26	43	51	37
	6.00	Eywtns News 6	59	55	56	58	57
	7.00	ABC Eve News	52	50	43	49	48
	Average		45	44	45	53	47
Overall Average			52	53	51	58	53

53%. Thus only about half the people who see one episode of a programme tend to see another. The table also shows that there is no substantial erosion of this repeat-viewing level for days further apart. There is a suggestion in Table 7.1 that repeat-viewing for *consecutive* days — Thursday and Friday — is a few percentage points higher, but more cases would be required to establish this as a firm finding. (We note that the Arbitron measurement week runs from Wednesday to Tuesday, so that the Monday and Tuesday results in this table refer to the extent to which the viewers then had seen the *previous* Friday's episode.)

There appears to be some fairly substantial variation about the overall average of 53% in the repeat-viewing levels for different programmes, and possibly for different networks. Repeat-viewing for the four programmes on WCBS (the New York CBS affiliate) are all in the 60's whereas the levels for the NBC and ABC afternoon films are rather low. But again substantially more study is needed to establish whether these variations are generalisable.

Repeat-viewing levels for the other New York stations studied — WNEW, WOR, WPIX — are mostly lower than for the networks. This is illustrated in Table 7.2 for WNEW, where strip-programming continues right through the evening. From 4 to 7 pm the repeat levels average at

Table 7.2

Repeat-viewing within the week: WNEW
(% of Monday, Tuesday and Thursday audiences of a programme
who watched the Friday episode of that programme)

New York Housewives Jan-Feb 74			% Viewing on Friday of the Audience on			
			Mon	Tue	Thu*	Av.
WNEW	4.00	Bugs Bunny	33	20	16	23
	4.30	Lost in Spce	31	33	42	35
	5.30	Flintstones	30	25	40	32
	6.00	Lucy Show	48	38	46	44
	6.30	Bewitched	39	38	39	39
	7.00	Mission Imposs.	35	42	40	39
	8.00	Dealers Choice	9	15	19	14
	8.30	Merv Griffin	30	33	34	32
	10.00	10 O'Clock Nws	43	40	40	41
	11.00	Step Beyond	24	29	26	26
	11.30	11.30 Movie	16	21	13	17
	Average		31	30	32	31

* Wednesday not tabulated

35%, compared with 53% in Table 7.1 for the major networks. Later in the evening the repeat levels are sometimes lower still, possibly due to the competition of the high-rating network programmes then.

Table 7.3 sets out the WNEW ratings over the week-days as background to this analysis. The ratings are mostly steady from day to day. But this is only on the surface — it does not mean that the audiences consist of the same people every day. As we have seen from Table 7.2, only about a third of those viewing a particular programme are repeat-viewers from one day to another. (The ratings are similarly steady for the networks — as was illustrated in Table 2.2 — but still only about half the viewers are the same from day to day.)

Table 7.3

Week-day ratings for WNEW programmes

New York Housewives Jan–Feb 74			% HW's Viewing				
			Mon	Tue	Thu*	Fri	Av.
WNEW	4.00	Bugs Bunny	1	1	1	1	1
	4.30	Lost in Spce	1	1	1	1	1
	5.30	Flintstones	1	2	1	1	1
	6.00	Lucy Show	2	3	2	2	2
	6.30	Bewitched	3	3	3	3	3
	7.00	Mission Imposs.	5	5	5	4	5
	8.00	Dealers Choice	3	1	2	1	2
	8.30	Merv Griffin	6	6	5	5	6
	10.00	10 O'Clock Nws	9	10	10	8	9
	11.00	Step Beyond	2	2	2	2	2
	11.30	11.30 Movie	1	1	1	2	1
	Average		3	3	3	3	3

* Wednesday not tabulated

Most of the ratings for WNEW are very low. This may help to explain the low repeat-viewing levels in Table 7.2. As noted in Chapter 5, repeat-viewing tends to decrease with rating in the UK. To what extent does rating level explain the repeat-viewing frequency in the US? WNEW's abnormally low 8 pm repeat-viewing in Table 7.2 goes with an unusually low 8 pm rating in Table 7.3. The low 4.30 repeat-levels for WNBC and WABC also go with relatively low ratings; and WCBS's high 4.30 pm repeat goes with a high rating. But these are hand-picked examples. Far more work is needed to establish valid and usable relationships in the US.

Another possible influence on repeat-viewing levels was also noted in

Chapter 5. Repeat-viewing analyses between two episodes are simple when the two ratings are steady, but problems arise when they are not. For example, for two episodes with ratings of 4 and 3, only 75% of the first audience could possibly see the second episode. And with low-rating programmes as for WNEW here, small differences in rating levels become proportionately large. Thus two ratings of 2 could reflect actual audience levels as different as 2·5 and 1·5, allowing a *maximum* possible repeat-viewing percentage of only 60%. With sample data, rating differences as such can arise simply because of sampling errors. The technicalities here require further study.

Duplication between different programmes

The level of audience overlap between different programmes in the New York data is usually lower than the level of repeat-viewing for different episodes of the same programme. Instead of averaging at 30 to 50%, audience duplication for different programmes on different days is generally below 20%. The low duplication levels are due to relatively low *rating* levels in New York where there are many channels. The duplication is still directly proportional to the programmes' ratings, as in the UK, and the duplication of viewing law of Chapter 2 again applies.

For the three national networks in the 1974 New York data, the duplication coefficient – the ratio of the duplicated audience to the rating – is mainly of the order of 1·5 or 1·6 for programmes on the same channel and about 1·1 or 1·2 for programmes on different channels. Thus the degree of channel loyalty (the within-channel factor of 1·5 or 1·6) is marginally smaller than that in the UK recently (1·7 or so). But the amount of switching between channels is relatively higher – 1·2 versus 0·9 in the UK.

Table 7.4 gives audience duplication figures for ABC programmes on Thursdays and Fridays – a typical example for a US network. The general tendency is for the Friday ABC programmes to be more popular among Thursday ABC viewers than they are in the population as a whole. Virtually all the Friday duplications in the table are higher than the Friday ratings, usually by a factor of 1·5 to 1·7.

For example, the 8 pm Friday programme (which was usually "Brady's Bunch" in the four weeks analysed) was seen by roughly 10% of the viewers on any ABC Thursday programme compared with its rating of 6, a duplication factor of just over 1·7. The 9 pm Friday programme ("The $6 Million Man" and others) was seen by about 20% of viewers of any

Table 7.4

Duplication of viewing between programmes on the same channel
(% of viewers of one programme who also watch another
on another day)

New York Housewives Jan-Feb 74			Who also viewed ABC on Friday at				
			7.30	8.00	9.00	11.00	11.30
Viewers of ABC on THURSDAY at							
4.00	Love Am Style-D	100%	17	13	23	23	10
4.30	4.30 Movie	100%	28	12	18	14	4
6.00	Eywtns News 6	100%	23	11	16	22	7
7.00	ABC Eve News	100%	31	12	19	23	7
7.30	Animal World	100%	20	7	12	15	3
8.00	Chopper One	100%	14	8	26	21	7
9.00	Kung Fu	100%	14	12	21	20	7
10.00	Strts Sn Frn	100%	13	8	24	24	6
11.00	Eywtns News 11	100%	16	9	21	(43)	(13)
11.30	Wide World Entert.	100%	16	11	18	(36)	(17)
	Average	100%	19	10	20	20*	6*
	1.6 x Rating		**19**	**10**	**21**	**18**	**6**
	Rating	100%	12	6	13	11	4

* Excluding late night cluster

Thursday programme compared with *its* rating of 13 — a duplication factor of 1·5. And so on.

The individual figures in each column of the table vary somewhat, but most of the differences from the column average are relatively small (on average about 3 percentage points) and they appear to be largely irregular. However, there are two exceptions, both having *higher* duplication.

The first exception is for the Friday 7.30 programme. It has relatively high duplications with the 7.30 pm or earlier programmes on the preceding day. This is a common finding for all week-days on the three networks. It is probably due to an "inheritance effect" from the 7 pm news, when the strip-programming type of high, day by day, repeat levels operate, as discussed earlier.

The second exception is the late-night cluster of high duplications, seen in the bottom right-hand corner of the table. Duplication between the audiences at 11 pm or later on each of the two days is markedly higher than shown by the rest of the table. This is again a general phenomenon on all networks in the US, and one we have also seen for the UK (Chapter 4). The cluster is thought to be due to many people habitually going to bed between 10 and 11, so that those who stay up tend to be the same people night after night. As argued in Chapter 4, it is the late-night ratings which are too low and hence make the duplications appear high. If late-night viewers were expressed as a percentage of those then *available* to view, the "ratings" would be higher. The late-night duplications would then no longer appear so high, but more in line with the patterns observed for the rest of the evening.

Direct numerical evidence of this point is lacking, but the duplication of the late-night viewers on one day for programmes at 10 pm or earlier on the next day does show that, as in the UK, there is nothing abnormal about the late-night viewers as such. They are not especially heavy viewers of television in general. If they were, their duplication levels would be high between different programmes generally. But it is only late in the evening that this happens.

The *general* tendency of audience duplication for the three main networks therefore is for viewers of one programme to be about 60% more likely to watch another programme on the same channel on another day than is the public in general, i.e. the percentage of duplicated viewers is about 1·6 times the rating. The duplication coefficients found so far differ slightly between the three network channels in New York, i.e. WCBS — 1·48, WNBC — 1·58, WABC — 1·64. To establish whether these small differences are generalisable over time and across the country would require further work. The striking feature is in any case the similarity rather than the difference of the three values.

The situation illustrated in Table 7.4 was for two consecutive days. But there is little change in the duplication coefficients for pairs of days further apart — at most a slight decrease. The average for the three networks is

consecutive days — 1·60
1 day apart   — 1·58
2 days apart   — 1·55
3 days apart   — 1·54

The trend is consistent but numerically very small. Far more extensive results would be required to establish its generalisability.

Comparable results about within-channel duplication for the three independent New York stations analysed here hardly occur because of the prevalence of strip-programming. Instead of duplication between different programmes at the same time on two days the situation is dominated by repeat-viewing of different episodes of the same programme. Repeat-viewing being generally much higher than normal duplication levels, the audience overlap for two programmes shown at different times is often still affected by the high repeat levels operating for programmes at the *same* times, through the inheritance effect (see below). Furthermore, individual results tend to be highly variable, because of small rating levels for the independent stations and, hence, small subsamples in the data. No simple conclusions are yet available.

Duplication between channels

Audience overlap between programmes shown on different days on *different* channels tends to be fractionally higher than the ratings, by a factor of 1·1 or perhaps 1·2. Thus viewers of an ABC programme are, if anything, very slightly more likely to see a CBS programme the next day than is the population as a whole.

Table 7.5 illustrates these cross-channel results, showing the percentages of viewers of ABC programmes on a Wednesday who saw the CBS programmes the following Tuesday. In this table, the duplications nearly equal the ratings, to within a few percentage points. But in other cases the duplications tend to be about 10 or 20% higher, giving the duplication coefficient of 1·1 or 1·2 mentioned earlier.

Table 7.5

Duplication of viewing BETWEEN channels

New York Housewives Jan-Feb 74			Who also viewed CBS on TUESDAY at				
			7.30	8.00	9.30	11.00	11.30
Viewers of ABC on WEDNESDAY at							
4.00	Love Am Style-D	100%	16	34	10	7	4
4:30	4.30 Movie	100%	14	22	11	2	3
6.00	Eywtns News 6	100%	11	26	13	3	3
7.00	ABC Eve News	100%	12	29	14	6	4
7.30	Strange Places	100%	12	29	10	5	2
8.00	Wed Mv of Wk	100%	8	24	12	7	3
10.00	Doc Elliot (etc.)	100%	13	28	12	6	5
11.00	Eywtns News 11	100%	10	24	15	5	3
11.30	Wide World Entert.	100%	9	23	14	7	2
	Average	100%	12	27	12	5	3
	Rating	100%	14	26	11	7	3

These findings — *within*-channel duplication coefficients of about 1·5 to 1·6 but between channel ones of about 1·1 — imply that channel loyalty exists in the main networks, but to a lesser degree than in the UK. This may be due to the greater similarity of the US network offerings, or perhaps to the stations' tendency to schedule programme changes at the same times (unlike ITV and BBC1 in the UK). (Between-channel duplications involving the smaller stations have not yet been effectively estimated because of statistical problems arising from the small samples of viewers that are involved.)

The inheritance effect

It is well known that for pairs of programmes shown on the same channel on the same day there is an "inheritance" effect, or "lead in" as it is called in the US. The audience duplication is larger because people stay tuned to the same channel when a programme ends, or tune in to the channel early to be sure of seeing a favourite programme there later (i.e. a "lead out" effect).

Table 7.6 shows typical same-day audience duplication figures for pairs of CBS Friday programmes. Except for some of the late-afternoon programmes and the 6 and 7 pm news shows, all but one of the overlap figures are less than 50%. Thus it is not a case of most people "being too lazy" to switch channels (or to switch off) once the programme they are viewing has finished.

Table 7.6

Same-day duplications: CBS on Friday

New York Housewives Jan-Feb 74			Who also Viewed CBS at								
			4.00	4.30	6.00	7.00	7.30	8.00	8.30	11.00	11.30
Viewers of CBS at											
4.00	Secret Storm	100%	100	60	51	38	13	14	13	9	8
4.30	Mike Douglas	100%	22	100	55	35	15	15	14	10	4
6.00	Ch 2 Nws - 6	100%	16	48	100	61	24	22	14	15	5
7.00	CBS Eve Nws	100%	13	32	64	100	34	18	15	15	5
7.30	Secrets Deep	100%	6	19	36	48	100	33	22	12	8
8.00	Dirty Sally	100%	5	16	26	21	27	100	40	15	7
8.30	CBS Fr Nt Mv	100%	4	12	14	14	15	33	100	24	8
11.00	Ch 2 Nws 11	100%	5	14	24	22	13	20	39	100	28
11.30	CBS Lt Movie	100%	8	12	17	16	17	19	27	56	100
Rating			4	12	14	13	9	11	14	9	4

However, the duplications are nearly all higher than the ratings shown at the bottom of the table. Thus viewers of one CBS programme are more likely than non-viewers to watch any *other* CBS programmes that day. But in most cases this is no higher than the normal positive audience duplication between two programmes on the same channel on *different* days. Table 7.7 allows for this general "channel loyalty" effect by subtracting the predicted between-day duplications.

Table 7.7

Lead-in effects
(differences between observed same-day duplications and
"between-day" estimates)

New York Housewives Jan-Feb 74	CBS Friday									Av.
	4.00	4.30	6.00	7.00	7.30	8.00	8.30	11.00	11.30	
CBS Friday										
4.00 Secret Storm	–	41	29	17	-2	-4	-9	-5	1	8
4.30 Mike Douglas	15	–	33	14	0	-3	-8	-4	-3	5
6.00 Ch 2 Nws - 6	9	29	–	40	9	4	-8	1	-2	10
7.00 CBS Eve Nws	6	13	42	–	19	0	-7	1	-2	9
7.30 Secrets Deep	-1	0	14	27	–	15	0	-2	1	7
8.00 Dirty Sally	-2	-3	4	0	12	–	18	1	0	4
8.30 CBS Fr Nt Mv	-3	-7	-8	-7	0	15	–	10	1	0
11.00 Ch 2 Nws 11	-2	-5	2	1	-2	2	17	–	21	4
11.30 CBS Lt Movie	1	-7	-5	-5	2	-3	5	42	–	4
Av. of adjacent programmes	15	35	37	33	15	15	18	26	21	24
Av. of adj. -but-one prog.	9	13	22	7	4	1	3	1	1	7
Av. all others	0	-4	-2	2	0	-2	-8	-2	-1	-2

As in the UK (Chapter 5), the results show three main features:

– the inheritance effect exists primarily between adjacent programmes;
-- for programme pairs separated by more than one other programme it is generally negligible, if it exists at all;
-- the effect varies in size (perhaps due to different programmes or times of day).

More work is needed to establish what factors determine the actual size of the inheritance effects and to differentiate between "lead out" and "lead in".

Summary

The results shown in this chapter indicate that many of the features of audience flow in the UK also occur in the US. The larger number of channels and the structuring into three main national networks, regional networks and a variety of independent stations, do not in themselves seem to lead to radically different viewing patterns. Repeat-viewing, audience duplication, channel loyalty and inheritance effects are largely as in the UK. There is relatively more switching among the three main networks, perhaps because their programme and scheduling policies are less distinct than those of the three channels in the UK.

The main difference with the UK is due to a particular aspect of programming policy, namely the high incidence of strip-programming within the week. This means that a particular channel will often have a much higher audience duplication from one day to the next because this is governed by the nature of repeat-viewing. Two episodes of a programme will have 30 to 60% of their audience in common, whereas duplication of viewing between different programmes is seldom greater than 20% in the US. (The duplication of viewing law still operates, but because of the larger number of channels in the US, rating levels for any one programme tend to be smaller than in the UK. Hence duplication tends to be numerically smaller as well, even though the duplication coefficients of the order of 1·6 or so for the main networks are roughly similar to the UK figures).

The amount of research into US audience flow reported here is still quite limited. More fleshing out of the initial findings is needed. The biggest gap so far is the systematic study of viewing of different episodes of the same programme in successive weeks. Given the findings so far, one might guess that such week by week repeat-viewing levels will tend to be about 55% — the same as for the UK and as for *within-week* repeat-viewing on major US networks. But some direct evidence is needed. Typical of such extensions of the UK findings, quite limited data should show whether or not the same pattern holds.

The studies so far carried out in the US have shown that despite differences in the TV scene, the approach to the analysis of audience flow developed in the UK is applicable to the US and tends to produce the same kind of simple and generalisable results. The implication is that the approach is also worth following in other countries (including ones where only one channel is available). Indeed even if audience flow patterns in some other countries should turn out to be very different, the present approach of examining repeat-viewing and audience duplication levels would help to pinpoint both the existence and the nature of such differences quickly and efficiently.

8 Audience Appreciation

So far in this book we have been concerned with people's viewing behaviour. That is a natural place to begin research into television. We need to know what people *do* before we can seek to explain their reasons for doing so or the effects of their actions. But information about audience size and patterns of viewing is perhaps not enough if we are to learn why people view particular programmes and how much pleasure or value they obtain from them.

Thus the level of repeat-viewing which a regular programme attracts need not be a complete guide to satisfaction. Repeat-viewing could be affected by other factors like the opposing programmes, family influence and the pull of the individual's other activities. What is more, the differences in the repeat-viewing levels of different programmes are generally not very large and are mainly related to rating level, as we have seen in Chapter 5. They therefore provide no effective guide to viewers' satisfaction in the aggregate.

Ratings vary a great deal more than do repeat-viewing levels but are obviously more a measure of mass appeal than of the individual viewer's satisfaction. They also can be greatly influenced by extraneous factors like time of day, competitive programming on the other channels, inheritance effects, and channel. For example, BBC2 still tends to do poorly in terms of audience sizes.

More direct indicators of viewers' appreciation of different programmes are therefore desirable. They should also shed light on more *detailed* aspects of audience reactions – a producer may wish to know which aspects of his programme people appreciated, how far the points made had been understood, and whether the programme achieved the objectives intended.

Programme producers and planning executives are not the only ones who require additional information to ratings. Critics of television have often attacked or defended it on grounds which have little to do with the viewing figures as such. The Pilkington Committee (1960), for example, had little time for ratings:

> It is by no means obvious that a vast audience watching television all the evening will derive a greater sense of enjoyment from it than will several small audiences each of which watches for part of the evening

only. For the first may barely tolerate what it sees: while the second might enjoy it intensely.

The Committee was interested in the "quality" of the programmes, their social and psychological effects and other aspects.

Even manufacturers and their advertising agencies, who generally receive most of the blame for the dominating influence of the ratings, require data to supplement ratings figures. They want to know about the attention paid to programmes, viewers' understanding, perception and recall of commercials, and how the advertising affects attitudes and the formation of intentions to buy.

Both the BBC and ITV interests therefore collect regular as well as *ad hoc* information on audience reactions. Most of the work refers to particular programmes at particular times and relatively few findings of general significance have yet been made. Many of the BBC's findings have been summarised in its first *Annual Review of BBC Audience Research Findings* (BBC, 1974) to which the reader is referred. In the present chapter we describe some general results which are emerging from the IBA's regular measurement of audience appreciation.

The Appreciation Index

The IBA makes regular measurements of audience appreciation on the basis of a panel, in alternate weeks about 500 adults in the London area reporting the programmes they have seen that week and their appreciation of them. In intervening weeks samples of viewers are covered in each of the other ITV regions in rotation.

For each programme seen in the particular week, the panel member marks his appreciation in terms of one of six categories. These run from "Not at all interesting and/or enjoyable" (scored 0), and increase by steps of 20 to "Extremely interesting and/or enjoyable" (scored 100). By averaging the scores of the different panel members, an average Appreciation Index (AI) running from 0 to 100 is obtained for each programme seen.

Viewers score only programmes which they have viewed that week and are unlikely to feel (or to *say* that they feel) that all of these have been a complete waste of time or positively unpleasant. As a result, average values mostly lie in the upper half of the range. Table 8.1 gives some typical examples of the distribution of individual scores and average AI values.

Table 8.1

Some percentage distributions of individual appreciation scores
and corresponding average AI values
(% of viewers giving scores)

London Adults 1972		Individual AI Scores						Average AI
		0	20	40	60	80	100	
Viewers of								
Pot Black	%	0	0	0	2	51	47	88
Colditz	%	1	0	0	16	41	41	84
Crossroads	%	0	1	3	26	34	36	80
News at Ten	%	0	0	3	27	43	26	78
Wrestling	%	8	0	1	25	33	34	75
Mr. Trimble	%	7	0	0	23	51	19	73
Coronation Street	%	3	6	6	23	34	24	71
Top of the Pops	%	8	4	8	41	27	11	61
Candid Camera	%	13	10	13	35	16	12	53
Play for Today	%	13	25	6	28	10	10	48

Appreciation index and audience size

In general there is no correlation between the AI and audience size (rating level). Table 8.2 illustrates this for a set of eleven programmes which featured in a particular analysis. Ratings (here the percentage of the AI panel who recorded having viewed the programme) vary from 37% down to 12%, but the AI scores show no parallel trend. In these cases they were generally close to the average of 76, with just two somewhat higher scores at 84.

The lack of a simple correlation here need not be surprising. The AI scores are given by those people who viewed the programme in question. There is no particular reason why those who watched a programme seen by few other people should like it less than those who saw a high-rating programme. On the other hand, it could have been that at least *some* low-rating programmes are seen by small audiences who enjoy them far more intensely than the general run of programmes (as suggested by the Pilkington Committee). Alternatively they might be seen by heavy viewers who do not actually like them very much but cannot switch off. We now know that neither of these extremes appear to be common.

There is some suggestion in the data analysed so far that highest AI scores tend to be given both to high and to low-rating programmes, with middle size audiences being less appreciative, but the statistical reliability and the numerical size of these differences is small.

Table 8.2

Audience size and AI score

London Adults 1972	Rating	AI Score
Man at the Top	37	84
Budgie	32	75
David Nixon Show	34	76
Smith Family	31	77
Star Trek	24	72
Saturday Variety	30	72
Parkinson	31	84
Man Outside	25	74
World in Action	19	76
This Week	18	77
Panorama	12	77
Average	27	76

Appreciation Index and repeat-viewing

Table 8.3 takes the same eleven programmes as in Table 8.2 and compares their AI scores with their repeat-viewing levels (i.e. the percentage of viewers in the AI panel in one week who watched the programme again two weeks later).

Table 8.3

Repeat-viewing and AI scores

London Adults 1972	Average Repeat %	Average AI Score
Man at the Top	76	84
Budgie	72	75
David Nixon Show	69	76
Smith Family	68	77
Star Trek	61	72
Saturday Variety	60	72
Parkinson	59	84
Man Outside	58	74
World in Action	50	76
This Week	46	77
Panorama	43	77
Average	60	76

The repeat-viewing level decreases systematically in the table. This clearly parallels the trend in the rating levels of these eleven programmes shown in Table 8.2 and is in line with the correlation between ratings and repeat-viewing in the JICTAR data noted in Chapter 5. But since the AI scores for the programmes in Table 8.3 hardly vary, there is no correlation between AI score and repeat-viewing here. This may seem surprising. More extensive checking is still required, but it looks as though the Appreciation Index does *not* measure the same kind of thing as the incidence of repeat-viewers.

Part of the explanation is probably that we are dealing with repetitive programmes. Someone who really does not greatly care for a particular programme will probably not even have seen the *first* of a pair of episodes that we are analysing, let alone the second.

Appreciation Index and individual repeat-viewing

More positive results arise when we dissect the audience of any single programme in terms of their individual appreciation and repeat-viewing. Taking "Man at the Top", the first of the eleven programmes just considered, we can break its audience down into the different Appreciation scores the viewers gave to the programme (grouping those scoring 0, 20, 40 and 60 together because of the small numbers of viewers giving these scores). We then find that there is a corresponding gradient in the incidence of repeat-viewing:

AI score	0–60	80	100
% repeat-viewers	62	72	85

Thus of those who gave a low AI score to this programme, only 62% saw it again next time (i.e. two weeks later in the particular biweekly situation in which AI measurements are made), whereas of those who appreciated the programme highly (an AI score of 100), 85% saw it again.

This pattern occurs not only for this programme, which had a rather high repeat-viewing level, but also for virtually all the other ten programmes in the analysis, as is shown in Table 8.4. (The only exceptions are the last two programmes, which had the lowest ratings, 18 and 12, so that sample numbers for the analysis in Table 8.4 are small and unreliable.)

These results indicate that the higher the AI score a person gives to a programme, the more likely he is to watch another episode of that programme. However, this is not a simple causal link. Someone giving a

Table 8.4

Repeat-viewing by AI score
(% of viewers giving the stated AI score who are repeat-viewers)

London Adults 1972	AI Score		
	0-60	80	100
Repeat-Viewers of	%	%	%
Man at the Top	62	72	85
Budgie	48	76	88
David Nixon	57	71	82
Smith Family	57	66	83
Star Trek	55	66	65
Saturday Variety	48	66	68
Parkinson	53	59	66
Man Outside	53	57	62
World in Action	44	54	57
This Week	48	42	48
Panorama	25	55	35
Average repeat-viewers	50	62	67

high AI score to a programme is also more likely to have seen *preceding* episodes of the programme.

This is shown in summary form in Table 8.5 which compares the average repeat-viewing percentages from the bottom row of Table 8.4 with the corresponding average percentage of viewers of a programme who had seen the episode two weeks *before*. The correlation is therefore between AI score and *frequency of viewing*.

Table 8.5

Seeing the preceding and succeeding episodes by AI score
(% of viewers giving the AI score who also saw
the preceding and succeeding episode)

Average of 11 Programmes 1972	AI Score Given to Current Episode		
	0-60	80	100
	%	%	%
Seeing previous episode	53	65	73
" succeeding "	50	62	67

Repeat, new and lapsed viewers

This correlation between AI scores and frequency of viewing can also be demonstrated by looking at the scores given to a programme by three different categories of viewers. Thus for any two episodes of a programme we have

"repeat-viewers", i.e. those who saw *both* episodes;
"new" viewers, i.e. those who saw the second but not first (although they may have seen yet earlier episodes);
"lapsed" viewers, i.e. those who saw the first but not the second episode.

Repeat-viewers of "Man at the Top" gave an average AI score of 87, but "new" and "lapsed" viewers liked the programme somewhat less, giving AI scores averaging in the mid-70's. This pattern holds for most of the eleven programmes, as shown in Table 8.6.

Table 8.6

Audience appreciation index by "repeat",
"new" and "lapsed" viewers

London Adults 1972	Type of Viewer		
	"Repeat"	"New"	"Lapsed"
AI Score for:			
Man at the Top	87	78	75
Budgie	80	61	60
David Nixon	79	67	70
Smith Family	81	69	69
Star Trek	75	73	62
Saturday Variety	75	67	71
Parkinson	86	84	83
Man Outisde	76	69	74
World in Action	77	75	71
This Week	77	73	76
Panorama	81	71	75
Average	79	72	71

The similarity of the average AI scores given by "new" and "lapsed" viewers confirms the results in Table 8.5 that a viewer's expressed appreciation of a particular episode of a programme does not seem to be more associated with his future viewing than with his past.

The general conclusion would therefore seem to be the rather unexciting one that the more enjoyable a person finds a programme the more often he is likely to watch it. However, the situation may not be quite that straightforward. Firstly, the effect may frequently be the other way round — the more often someone watches a programme, the more enjoyable he finds it. Secondly, it appears that heavy viewers of television generally tend to give somewhat higher AI scores than do light viewers. But repeat-viewers of a particular programme usually include an above-normal proportion of heavy TV viewers. So the heavy viewer's higher appreciation, rather than his reaction to the particular programme, may be the explanation of the above findings. A good deal of further detailed research is needed here.

"Coronation Street": a case history

An example of how these kinds of results can already help to explain practical findings is provided by results for the twice-weekly ITV programme "Coronation Street".

In a certain analysis over several months in 1972 it was found that the AI scores for "Coronation Street" were consistently higher on Wednesdays than on Mondays:

> average Monday AI   70
> average Wednesday AI   73

Why did two adjacent episodes of the same long-running programme register such a consistent (if small) difference in audience appreciation?

The situation appears at first sight more curious when we take audience size into account. The Monday episodes were seen consistently by fractionally more people than the Wednesday ones:

> average Monday rating   40
> average Wednesday rating   38

Thus despite somewhat lower AI scores given to the Monday episodes, slightly more people saw them than those on Wednesdays.

The explanation lies in the fact that some viewers saw *both* "Coronation Street" episodes in a given week ("repeat-viewers") and some only one ("new" or "lapsed" viewers). In line with the more general findings illustrated in Table 8.6, Monday–Wednesday repeat-viewers gave higher AI scores than the others, the difference being here as much as 15 AI points:

average AI score of repeat-viewers	75
average AI score of one episode only viewers	60

The repeat-viewers gave almost the same average scores to the Monday and Wednesday episodes, and by definition, the *number* of repeat-viewers is the same on Monday and on Wednesday – they are the same people. But since the rating of the Monday episode was higher than that of the Wednesday one, there were more one episode only viewers on Monday, i.e. more of those who tend to give the programme lower AI scores. The Monday audience therefore contained more relatively low AI-scoring viewers than the Wednesday audience. The difference in the AI scores of the Monday and Wednesday episodes was therefore not due to any difference in the intrinsic merit of the episodes, nor how the same people evaluated them. Instead it was due to whatever other factors (e.g. viewing habits, opposing programmes, etc.) caused the Monday episode to be seen by slightly more viewers than that shown on Wednesday.

Programmes of the same type

Another illustration of the discriminative ability of the Appreciation Index was apparently provided by a case where those who gave a particular programme a high score tended to be more likely to watch a subsequent episode and also other programmes of the same type. This however has turned out to be a warning against rushing to general conclusions without first establishing the generalisability of a new result.

Table 8.7 shows how at a certain point in 1972, those viewers of the ITV current affairs programme "This Week" who scored it 80–100 contained somewhat higher proportions watching another edition of the same programme or other similar programmes, like "World in Action" and "Panorama", than those who scored "This Week" only 0–60.

Table 8.7

Some viewing patterns according to AI score for "This Week"

London Adults 1972	AI Scores for a given edition of "This Week"	
	0-60	80-100
% also Viewing	%	%
This Week (another edition)	45	55
World in Action	30	53
Panorama	8	13

Even when directly competitive channel switching is required, the same type of effect occurred. For example, "World in Action" and "Panorama" were screened at the same time (Monday 8 pm) on two different channels (ITV and BBC1), with an entertainment programme being shown on BBC2. Table 8.8 shows how more of those who liked "Panorama" one week (AI scores of 80–100) watched it again the following week, in line with the preceding results. And of those high scorers who did *not* actually view "Panorama" again, a relatively high proportion (21%) saw "World in Action", the competing current affairs programme, instead.

Table 8.8

Viewing patterns at the same time in a later week according to AI score for "Panorama"

London Adults 1972	AI Scores for "Panorama" in a given week (8 pm Monday BBC 1)	
	0-60	80-100
% who at 8 pm on Monday Two Weeks Later Viewed	%	%
Panorama (BBC1)	31	53
World in Action (ITV)	8	21
Neither	61	26
Total	100	100

As a consequence, of those who liked the earlier "Panorama" (AI scores of 80–100), only 26% did not see either current affairs programme two weeks later. In contrast, as many as 61% of the viewers of "Panorama" on a typical Monday at 8 pm who did *not* greatly like it (AI scores of 0–60) did *not* see a current affairs programme on Monday at 8 pm two weeks later, whether "Panorama" itself on the same channel (BBC1) or "World in Action" on another channel (ITV).

Here then we appear to have a certain clustering effect for different programmes of the same type — those who *like* them watch them more. *However, subsequent work on AI data in 1974 has so far largely failed to reproduce these particular results for current affairs programmes.*

Summary

The level of viewers' expressed "appreciation" of the programmes they

106

have seen does not vary with the size of the audience. Large-rating programmes are not more popular with their viewers than are small-rating programmes with theirs.

A programme with small audiences is generally seen either by people who specially like it, or merely by heavy viewers who have to watch *some* programme without necessarily liking it at all.

Appreciation scores do not correlate with the level of repeat-viewing for different programmes either. This is probably because a programme's overall repeat-viewing level does not itself appear to imply any special liking or disliking of the programme.

Frequent viewers of a repetitive programme do however tend to give higher appreciation scores than infrequent viewers. But this may be at least in part because heavy viewers of television generally "like" all programmes more.

It is clear from the various cases discussed in this chapter that viewers' attitudes towards television programming can only be properly interpreted in the context of their actual viewing behaviour.

9 The Liking of Programme Types

The studies of viewing behaviour summarised in earlier chapters have consistently shown that there is little or no "programme type" effect, whereby different programmes of the same type attract the same group of viewers. Nonetheless, one feels that two adventure series programmes, like "Hawaii 5-0" and "The Persuaders", should have some common appeal. Some people must like various programmes of this sort, some must prefer light entertainment programmes, others sport, and so on.

In fact there is evidence that this is so in terms of the programmes people say they like to watch. But this occurrence of programme-type clusters in terms of what people say they like can also be reconciled with the absence of such clusters in terms of what people actually *do*.

The data analysed here come from a question in the 1972 Leo Burnett *Life Style Research* study (see Segnit and Broadbent, 1971). Some 7000 adults were asked to say for each of 55 ITV or BBC programmes whether they:

"Really like to watch it";
"Watch it only because someone in my family likes it";
"Watch it when there's nothing better"; or
"Don't watch it".

It is important to note that "watching" a programme here refers to some general tendency to watch that programme *sometimes*, and not necessarily to having watched it on its last screening.

Programme clusters

To illustrate, we start with two marked groupings or clusters — sports programmes and current affairs programmes — which show up in terms of what people said they "really like to watch". For simplicity we shall refer to this as "liking".

Table 9.1 shows the percentage of adult viewers who "liked" one sports programme who also "liked" each of the other four sports programmes analysed. The table format is that of the duplication of viewing tables in earlier chapters, and the results are also similar. Thus there is little variation of the figures in each column from their average. "World of Sport" was liked by about 76% of those who liked any of the other sports programmes, and "Rugby Special" was liked by about 30%.

Table 9.1

Sports programmes
(% of adults saying they "really like to watch" one programme
who also say they "really like to watch" another)

UK Adults 1972		Who also like to watch:				
		World of Sport	Match of the Day	Grand-stand	Prof. Boxing	Rugby Special
Adults who like to watch:						
ITV World of Sport	100%	-	73	72	61	28
BBC Match of the Day	100%	75	-	71	60	30
BBC Grandstand	100%	80	77	-	62	32
ITV Prof. Boxing	100%	75	72	68	-	31
BBC Rugby Special	100%	74	75	75	65	-
Average	100%	76	74	72	62	30
ALL ADULTS	100%	39	38	35	32	15

These averages are all about twice as high as the percentages among adults as a whole who "like" the programmes. This 2:1 ratio applies for each separate programme.

Table 9.2

Current affairs programmes

UK Adults 1972		Who also like to watch:				
		Pano-rama	24 Hours	This Week	To-day	Line Up
Adults who like to watch:						
BBC Panorama	100%	-	68	50	37	18
BBC 24 Hours	100%	67	-	53	39	20
ITV This Week	100%	58	61	-	43	19
ITV Today	100%	47	50	48	-	17
BBC Line-Up	100%	59	66	53	44	-
Average	100%	58	61	51	41	19
ALL ADULTS	100%	31	30	27	24	9

Table 9.2 shows a similar pattern for five current affairs programmes. The percentage liking a programme amongst "likers" of another current affairs programme is again almost double the percentage liking the programme in the population as a whole.

These high "likings" could merely represent a general tendency amongst "likers" of one programme to "like" other programmes, irrespective of their type. Table 9.3 shows that this is partly so but not the full explanation. The current affairs programmes were liked substantially less often by likers of the sports programmes. There are real groupings here by programme type.

Table 9.3

Sports versus current affairs

UK Adults 1972		Who also like to watch:				
		Pano-rama	24 Hours	This Week	To-day	Line Up
Adults who like to watch:						
World of Sport	100%	41	39	34	29	12
March of the Day	100%	38	38	31	26	11
Grandstand	100%	42	40	34	28	12
Prof. Boxing	100%	42	39	34	28	13
Rugby Special	100%	47	44	33	29	16
Average	100%	42	40	33	28	13
ALL ADULTS	**100%**	**31**	**30**	**27**	**24**	**9**

Table 9.4 summarises these results. In the last column we see that the average current affairs programme is liked by 46% of the likers of the other current affairs programme and by only 31% of the likers of the sports programmes. But among adults as a whole the figure is even lower, at 24%.

Table 9.4

Likers of the current affairs programmes

UK Adults 1972	% who like to watch:					
	Pano-rama	24 Hours	This Week	To-Day	Line Up	Average prog
Amongst likers of the						
- OTHER C.A. progs.	58	61	51	41	19	46
- Sports programmes	42	40	33	28	13	31
ALL ADULTS	**31**	**30**	**27**	**24**	**9**	**24**

There is therefore some tendency for likers of one type of programme to also have a more than average liking for a programme of another type (the 31% versus 24% of the previous paragraph). This occurs generally. It seems to reflect the existence of heavier viewers who tend to watch – and often to "like" – a wide range of programmes. There certainly is almost no case of fewer "likers" of one programme type "liking" another (i.e. that *fewer* like it than occurs among all adults). The special clusters of more intense correlations for programmes of the same type illustrated in Tables 9.1 and 9.2 are therefore superimposed on this general tendency for some people to like TV programmes generally.

Table 9.5 gives another illustration for five light entertainment programmes on ITV, contrasted with adventure series and other ITV programmes. The contrast is clear.

Table 9.5

Liking of light entertainment on ITV

UK Adults 1972	Average % who like to watch:					
	Oppor-tunity Knocks	Family at War	Coro-nation Street	Golden Shot	Peyton Place	Ave-rage
Amongst likers of the						
- OTHER L. E. progs.	68	61	59	49	31	54
- Adventure Series	52	51	44	37	26	42
- OTHER ITV progs.	52	45	40	38	21	39
ALL ADULTS	**42**	**40**	**34**	**28**	**18**	**32**

Six programme clusters

Altogether, only six recognisable programme clusters have emerged from the data analysed so far (55 programmes covered in the Leo Burnett survey). They are summarised, with typical 1972 programmes, in Table 9.6.

It is important to stress that these clusters have been defined by noting the cases where people who say they "really like to watch" one particular programme include an especially high proportion of people who say they "really like to watch" another. The groupings have not come from any direct assessment of programme content or treatment. Nonetheless, five of the clusters agree with "common-sense" programme-type classifications (as also formalised by the IBA for example). These clusters can therefore

112

Table 9.6

Six programme clusters
(The six main programme groups emerging from the analysis)

Programme Cluster	Examples on ITV	Examples on BBC
1. SPORTS	World of Sport Professional Boxing	Grandstand Match of the Day Rugby Special
2. CURRENT AFFAIRS	Today This Week Aquarius	Late Night Line-Up Talk Back 24 Hours Panorama
3. LIGHT ENTERTAINMENT (a) Serials (b) General (c) Sit.-Comedy	(a) Coronation St. Peyton Place Family at War (b) Opportunity Knocks Golden Shot This is Your Life Mike & Bernie (c) Please Sir On the Buses	 (b) Z-Cars Owen M.D. Galloping Gourmet (c) Now Take My Wife Here's Lucy
4. ADVENTURE	Public Eye Callan Jason King Hawaii 5-0	Ironside The Virginian
5. CHILDREN'S		Magic Roundabout Blue Peter
6. (Not Named)	Thunderbirds	Star Trek Pink Panther Monty Python Top of the Pops

be readily named, as is done in Table 9.6. Only the sixth cluster does not fit into an existing category. (One can imagine these programmes being liked by the same people, but it is not clear how to say in standard terms what the programmes have in common; Leo Burnett refers to such programmes as "cult" programmes.)

113

The typology in Table 9.6 should not be overinterpreted. It does not mean that viewers can be divided into separate sub-groups who like one type of programme and *not* others. It is not an exclusive classification in that sense. Instead, it only reflects groupings of certain above-average likings.

Exceptions

The analysis here is based on only one set of programmes at one particular point in time (1972) and further work is needed. (Corresponding analyses of the AI type of data which was described in Chapter 8 are leading to broadly similar results. The AI data are however complicated for this particular purpose by the assessments being related to the programmes actually seen that week.)

Some exceptions to the main clusters are worth noting at this early stage. A few programmes among the fifty-five, like the highly popular comedy show "Morecambe and Wise", could not readily be assigned to *any* programme cluster. One or two other programmes fell into more than one group. "Wrestling" on ITV is one such example. As shown in Table 9.7, it is popular among those who said they like to watch ITV's "World of Sport" and "Boxing". But it is less popular among those who say they liked BBC's sports programme "Match of the Day" and "Rugby Special". In fact, "Wrestling" is *more* popular with people who said they like entertainment programmes like "Golden Shot" or "Opportunity Knocks".

Table 9.7

Wrestling

(% of adults saying they "really like to watch" each programme who also say they "really like to watch" wrestling)

UK Adults 1972		Who also like to watch: Wrestling
Adults who like to watch:		
World of Sport	100%	48) 50
Prof. Boxing	100%	51
Match of the Day	100%	40)
Grandstand	100%	43) 40
Rugby Special	100%	37)
Golden Shot	100%	49) 47
Opportunity Knocks	100%	46
ALL ADULTS	100%	32

"Wrestling" therefore appears to appeal to two somewhat different groups of people for two different reasons — to one it is sport, to the other light entertainment.

Programme character

These data on the programmes which viewers say they really like to watch can be used to define the appeal or "character" of a programme. This can be done by noting how popular the programme is among the people who like various *other* programmes. The approach is illustrated in Table 9.8 for "Family at War".

Table 9.8

Programme character: "Family at War"
(10 programmes whose "likers" most like, or
least like, "Family at War")

UK Adults 1972	% who say they like to watch FAMILY AT WAR
ALL ADULTS	**40**
Among adults who like:	High
Peyton Place	72
Coronation Street	67
This is Your Life	58
Golden Shot	55
Opportunity Knocks	54
Mike & Bernie	52
Here's Lucy	52
Owen M.D.	52
Public Eye	52
Jason King	53
	Low
Pink Panther	32
Monty Python	30
Magic Roundabout	32
Panorama	33
Tomorrow's World	31
World of Sport	33
Grandstand	30
Match of the Day	29
Prof. Boxing	28
Rugby Special	26

From the fifty-four other programmes covered, the ten programmes are shown whose fans include the highest percentage of people who also like "Family at War", and the ten programmes which have the *least* likers in common with it. Thus across the population as a whole, 40% say they like "Family at War". But people who like *other* programmes show a great deal of variation in this respect, from a high of 72% to a low of 26%.

At one extreme, "Family at War" is very popular with people who like the other two ITV serials, "Peyton Place" and "Coronation Street" (being liked by about 70% of these). It is also relatively popular with people who said that they like other light entertainment programmes, and with those who said they like adventure series such as "Public Eye" and "Jason King".

At the other extreme, relatively few of those who said they like current affairs, sports, or the "Monty Python" group of programmes (cluster 6 in Table 9.6) also say they like "Family at War". Here the incidence of people saying they like "Family at War" is actually well below the level among all adults — one of the quite rare instances of this in the current data. The result for "Family at War" is therefore a wide spectrum of opinions.

Table 9.9 gives a slightly more complicated example, for "Aquarius". The general level of "liking" is much lower (only 12% of adults as a whole do so). "Aquarius" is relatively popular among those who said they like some of the current affairs programmes, especially "Line Up", but not particularly among fans of other programmes in the group such as "Panorama". "Aquarius" also appeals to people who said they like the "Monty Python" and adventure types of programmes. However, the programme is relatively unpopular with both extremes of the dimension seen for "Family at War" — the fans of "This is Your Life" or "Coronation Street" and the fans of the sports programmes.

The usefulness of such analyses will depend on experience of a wide range of examples and the development of a "feel" for different patterns or preferences. In particular, it will be possible to differentiate in detailed terms between some programmes which are in most respects similar. For instance, "Talk Back" and "Line Up" had a particularly high overlap in terms of the people who said they like them. Both were popular among people who said they like other current affairs programmes. But "Talk Back" was also relatively popular with those who say they like some of the lighter programmes, such as "Galloping Gourmet" or "Owen M.D." while "Line Up' was popular with those who said they like more "intellectual" programmes such as "Aquarius" and "Monty Python".

Table 9.9

Programme character: "Aquarius"
(10 programmes whose "likers" most like, or least like, "Aquarius")

UK Adults 1972	% who say they like to watch AQUARIUS	
ALL ADULTS	**12**	
Among adults who like:		High
Line Up	28	
This Week	20	
Parkinson	17	
Braden's Week	. 17	
Monty Python	19	
Thunderbirds	19	
Star Trek	17	
Jason King	17	
Callan	17	
The X-Film	17	
		Low
Owen M.D.	12	
This is Your Life	11	
Coronation Street	11	
Val Doonican	11	
Golden Shot	11	
Generation Game	11	
World of Sport	12	
Grandstand	11	
Rugby Special	11	
Match of the Day	10	

The nature of programme clusters

The programme clusters described in Table 9.6 are not unexpected on common-sense grounds. One *expects* certain people to like programmes of certain types. Nonetheless, the nature of the clusters does not necessarily turn out to be self-evident when we analyse them further.

Sports and light entertainment programmes are generally known to have markedly different appeal to two different groups, men and women. Thus the five sports programmes in Table 9.1 are typically more popular with men, almost 50% of whom said they "like" these programmes compared with only 16% of women. In contrast, light entertainment programmes are generally more popular among women, 42% of whom

117

"liked" the average programme in Table 9.5 compared with only 22% of men.

Thus it might be thought that the sports cluster would be due to men (who like sports programmes) and the light entertainment cluster due to women. But the reverse is the case. This can be seen from Table 9.10 which shows separately for women and for men the percentages "liking" sports programmes among those liking the other four sports programmes, together with the percentage of all women or all men who "like" each of the sports programmes.

Table 9.10

Sports: women and men
(% of women or men saying they "really like to watch"
one sports programme who also "really"
like to watch" other sports programmes)

UK 1972		Who also like to watch:				
		World of Sport	Match of the Day	Grand- stand	Prof. Boxing	Rugby Special
WOMEN who like to watch:						
World of Sport	100%	-	57	57	35	23
Match of the Day	100%	67	-	57	33	24
Grandstand	100%	72	61	-	32	27
Prof. Boxing	100%	66	52	48	-	27
Rugby Special	100%	65	58	61	41	-
Av. Sports Progr.	100%	68	57	56	35	25
ALL WOMEN	100%	23	29	18	12	8
MEN who like to watch:						
World of Sport	100%	-	80	78	73	31
Match of the Day	100%	78	-	76	70	32
Grandstand	100%	83	83	-	72	34
Prof. Boxing	100%	77	77	73	-	32
Rugby Special	100%	77	81	81	75	-
Av. Sports Progr.	100%	79	80	77	73	32
ALL MEN	100%	57	59	54	54	23

For women, there is a ratio of about 3:1 for each sports programme between likers of *other* sports programmes and All Women. In contrast, the corresponding differences for men are much smaller; the ratio of the figures in the last two rows of Table 9.10 is only about 1·4:1.

It follows that there is only a small group of women who like sports programmes, *that* is where the clustering of sports programmes among all adults mainly stems from. The men themselves only show relatively weak clustering (since so many like the sports programmes anyway).

The same type of pattern holds for light entertainment programmes, but with the role of men and women reversed. This is summarised in Table 9.11. Here there is relatively little clustering of the programmes among women, the differences between "likers" and all women being relatively small (the averages of 59% and 44%, a ratio of only 1·3:1). In contrast, only an average of 22% of *all* men "liked" the ITV light entertainment programmes whereas 43% of the male *"likers"* did so — nearly twice as many.

Table 9.11

Light entertainment: women and men
(the average % of women or men saying they "really like to watch"
one light entertainment programme who also say they
"really like to watch" another)

UK 1972		Who also like to watch:					
		Oppor- tunity Knocks	Family at War	Coro- nation Street	Golden Shot	Peyton Place	Ave- rage
WOMEN who like to watch:							
Av. OTHER L. E. progr.	100%	68	70	66	51	40	59
ALL WOMEN	100%	46	55	46	44	28	44
MEN who like to watch:							
Av. OTHER L. E. progr.	100%	67	43	47	46	14	43
ALL MEN	100%	37	23	21	24	7	22

These results indicate that the apparent clustering of programme preferences noted in this chapter is more a negative than a positive effect. It may reflect the existence of a group who do *not* like a particular group of programmes. More work is needed in this area.

Reconciliation with viewing behaviour

Whatever its underlying nature, the clustering of viewers' liking for programmes of the same type seems at first sight to contradict the finding that no such programme-type clusters have emerged for people's actual *viewing* behaviour (Chapter 4). But in fact the two kinds of results are not inconsistent.

People often watch programmes which they would not claim to like particularly. Apart from mentioning programmes which they "really like to watch", people in the Leo Burnett survey analysed in this chapter were also asked to mention programmes which they watched because "someone in the family likes them" or because "there's nothing better to watch". For example, 59% of adults claimed to watch "Panorama" at all. Of these, 31% said they really liked to watch it, 8% said they watched it because someone in the family liked it, and 15% said they watched it because there was nothing better to watch. The "enforced" viewing of "Panorama" by 23% out of the 59% who view serves to dilute the effect of programme preferences on actual viewing behaviour.

This dilution is illustrated in Table 9.12 for the current affairs versus sports programmes analysed earlier. The top part of the table repeats from Table 9.4 the average percentages who "really liked to watch" each current affairs programme, both among those people who "really liked to

Table 9.12

Preference versus viewing behaviour: current affairs
(% who either say they "really like to watch" or
"view for any reason" each current affairs programme among
those who also say they really like to watch or view for any reason:
OTHER current affairs — averaged over 4 programmes; and
sports — averaged over 5 programmes)

UK Adults 1972	Who also "really like to watch":					
	Pano-rama	24 Hours	This Week	To-day	Line Up	Av. C.A.
Adults who really like to watch :						
Av. OTHER C.A. 100%	58	61	51	41	19	46
Av. Sports 100%	42	40	33	28	13	31

	Who also "watch for any reason":					
	Pano-rama	24 Hours	This Week	To-day	Line Up	Av. C.A.
Adults who watch for any reason :						
Av. OTHER C.A. 100%	74	78	67	51	32	60
Av. Sport 100%	64	66	58	45	28	52

watch" another current affairs programme (an average of 46%) and among those who "really liked to watch" a sports programme (an average of 31%). The ratio is about 1·5:1.

The bottom part of the table gives comparable figures relating to people's *total* viewing claims, i.e. including "enforced" viewing of both kinds. It shows the average extent to which each current affairs programme was said to be watched for *any* reason by the people who for any reason watched another current affairs programme (an average of 60%) and by those who for any reason watched a sports programme (an average of 52%). The figure for duplicated viewing *within* the current affairs group is still somewhat the higher, but only by relatively little – an average ratio of just over 1·1:1. The clustering of different current affairs programmes is therefore very much weaker in terms of something which resembles actual measured viewing (i.e. "watching for any reason") rather than claims to *like* watching.

This is a general finding. Table 9.13 sets out the corresponding (but fuller) data for total viewing claims for two other programme groups, light entertainment and adventure series. A light entertainment programme is

Table 9.13

ITV light entertainment and adventure series: viewing for any reason
(% of adults claiming to view one programme for any reason
who also claim to view another for any reason)

UK Adults 1972		Who also view (for any reason):				
		Opportunity Knocks	Family at War	Coronation Street	Golden Shot	Peyton Place
Adults who view (for any reason):						
Opportunity Knocks	100%	-	68	71	82	39
Family at War	100%	79	-	72	77	42
Coronation Street	100%	86	75	-	81	43
Golden Shot	100%	82	66	67	-	37
Peyton Place	100%	85	80	79	83	-
Av. LIGHT ENTERTAINMENT		83	72	72	80	40
Thunderbirds	100%	79	70	66	80	45
Hawaii 5-0	100%	76	70	65	76	40
Jason King	100%	80	73	71	79	43
Persuaders	100%	76	69	65	75	38
Callan	100%	75	70	65	75	40
Public Eye	100%	78	73	68	77	41
Av. ADVENTURE SERIES		76	71	66	77	41

watched (for any reason) by on average 69% of those who watched (for any reason) the other light entertainments, and by on average 66% of those who watched (for any reason) an adventure series — the figures are almost the same. The difference which typically occurred for "liking" claims (e.g. an average of 54% among "likers" of the light entertainment programmes versus 42% among likers of the average adventure programme in Table 9.5) has virtually disappeared here in terms of people's general viewing behaviour.

An additional factor in reconciling the liking and viewing results is that in any particular week some people tend to miss seeing even those programmes which they particularly like. Viewing claims made in the context of the Leo Burnett survey analysed here are claims to view a given programme *sometimes*. The total claims to watch a programme "for any reason", e.g. the 59% claim for "Panorama", are well above the actual ratings normally achieved by any single showing of the programme. This is in line with what we already know about the irregularity of viewing of a programme series. Only just over half of those who see one given episode of a programme also see another given episode of it; hence many more see it *sometimes* during a period of weeks than see a single episode.

This irregularity of actual viewing behaviour (together with a certain amount of false claiming which can occur in an interview situation) could appear sufficient to complete the reconciliation of quite marked programme-type patterns in claimed viewing preferences with the lack of such patterns in the general viewing data.

Summary

Programmes fall into a recognisable classification in terms of what people say they "really like to watch". From some early analyses five named programme types have emerged so far: sports, current affairs, light entertainment, adventure and children's programmes. Programmes fall into such groups in as far as the people who say they like one programme in the group include an especially high proportion of people who also say they like other programmes in the same group. This classification of programmes is not exclusive: people who say they particularly like to watch one type of programme do not, in general, say they dislike other types.

However, the way in which these programme-type clusters appear to arise is perhaps unexpected. The underlying factor seems to be the existence of a group of people who do *not* like a particular type of programme.

The programme-type patterns in terms of claimed programme preferences have not been found in studies of actual viewing behaviour. This is partly due to the fact that a considerable part of a person's viewing is "enforced" — watching programmes because other family members want to see them or simply because there is nothing better to watch — and partly because of irregular viewing (only about 55% of the audience to one episode of a programme will watch the same programme in another week). Many people do not watch regularly even those programmes which they say they really like to watch.

10   Television as a Medium

In this final chapter we shall draw together the findings described in this book and briefly comment on some of the broader implications for television as a medium.

Simple findings

The results in this book are simple. There is one main result concerning the repeat-viewing of a given programme, one main result concerning the duplication between the audiences of different programmes, and some findings concerning the audience's expressed appreciation or liking of a programme.

Repeat-viewing

The basic finding here is that about 55% of viewers of one episode of a programme also watch the following episode (usually a week later). This single figure — 55% — summarises the main result succinctly.

Repeat-viewing levels, however, vary with audience size — the larger the audience, the higher the incidence of repeat-viewers. But otherwise, repeat-viewing levels of different programmes vary relatively little. (Exceptions occur, but are exceptional). In particular, there is virtually no systematic variation by programme type or content. Repeat-viewing of a serial with a continuing story-line is generally no higher than that for a regular film slot with radically different showings each week.

Nor is there evidence of any marked "erosion" in the degree of repeat-viewing for episodes shown more than one week apart. Failure to repeat-view seems to be a reflection of irregular or infrequent viewing habits, not of any special dislike or lack of interest in what has already been seen. This conclusion is supported by the finding that repeat-viewing is about 50% even for irregular programming, i.e. when altogether different programmes (with similar ratings) are shown in the same time-slot each week. Repeat-viewing therefore appears to be more a function of social habits (i.e. people's availability) than of programme content.

Extension of the analyses to more than two episodes of a programme also shows regular results and the development of *theory*. For any extended series of episodes the main finding is that almost no one sees all, or even nearly all, the different episodes.

Audience duplication

The extent to which different programmes share the same viewers also follows a simple pattern. This is expressed by the duplication of viewing law. Thus for any two programmes the level of duplication in their audiences depends on the *ratings* of the programmes and not on their *content*. One pair of programmes generally has the same degree of audience duplication as any other pair of programmes with the same ratings.

There are no subpatterns in this by programme type, but there are differences by channel. Audience duplication is higher for two programmes on the same channel than for ones on different channels. This is "channel loyalty" — some people are consistently heavier viewers of one channel than another.

The theoretical explanation of the duplication of viewing law is in terms of people's differing patterns of viewing and programme preference. One person watches programmes A, B and C, another watches A, X and Y, a third B, Y and M, and so on. It is not a case of there being large subgroups of people with common viewing patterns (other than is reflected in the sheer audience size for different programmes — the ratings). The observed patterns of audience duplication derive essentially from people's individualistic use of the medium, coupled with marked differences in their total amounts of viewing.

For two programmes both shown in the afternoon or in the early evening of two different week-days, audience duplication levels are much higher in relation to the ratings of these programmes. This is largely a matter of *non-availability*. The same people tend to be out on different week-day afternoons. Audience overlap is therefore high compared with the relatively low ratings in the population as a whole. The same occurs for pairs of programmes shown *late* on two different evenings.

Another consistent subpattern is that successive programmes on a channel have higher duplicated audiences. But this "inheritance effect" applies only to the adjacent programme and, to a lesser extent, to the next but one programme.

126

Audience appreciation

When people are asked about their liking or appreciation of programmes viewed in a given week, viewers' average "appreciation score" does not depend on the rating of the programme or on the incidence of repeat-viewing. But people giving a high appreciation score to a particular programme are more likely to see it often.

When people say they "really like to watch" a certain programme, they tend to say this also about other programmes of the same type (e.g. sport). This clustering of programmes people say they like to watch reconciles with the *lack* of such programme-type preferences in actual viewing behaviour, partly because people do not always watch even programmes they really like and partly because they also watch programmes other than ones they "really like to watch" (i.e. if "someone else in the family likes it", or if "nothing better was on").

Some broad implications

The main implications of these various findings is that television as currently operated is indeed a mass medium. Instead of being complex. with much differentiation between distinct groups of viewers or between the audiences of different programmes, viewing behaviour and audience appreciation appear to follow a few very general and simple patterns operating right across the board.

For example, it has been long established that the AB social classes in the UK view somewhat less television than does the public as a whole. None the less, the actual distributions of viewing times between different programmes are similar. This is so even though preference is expressed by the AB's for more "serious" material (e.g. Marplan, 1965). But there are discrepancies between what viewers say or feel they would like to watch and what they watch in practice. A similar conclusion was reached in Steiner's (1963) study of viewing tastes and behaviour in the US, reinforced by a recent follow-up (Bower, 1973).

The patterns of viewing behaviour established in this book cannot in themselves determine the decisions that ought to be made, either about television in general or about particular programmes. For decisions, certain social or other criteria or targets need to be set. But in as far as these refer to the viewer, such criteria can now be informed by, or evaluated against, our knowledge of viewer behaviour. (For example, we now know that it would be absurd to set a target of 90% repeat-viewers for successive episodes of a new series — that sort of thing simply does not happen. And if we observe a repeat-level of about 50 to 60%, we now know that this is normal. It also follows that we cannot expect many people to see all or most episodes in any programme series. The medium does not work like that.)

The results in this book do not in themselves say for example whether or not there should be a fourth channel in the UK, how it ought to be organised, or what sort of programming policy ought to be adopted. Some kind of target – like "increasing viewers' choice of programmes" – needs to be set. But we can now consider whether provision of an additional channel would really result in increased choice, or what such a target means anyway. Viewers may not use the additional "freedom of choice" appropriately.

Thus the evidence is that most viewers have a wide spread of interests and desires as far as their viewing is concerned. This suggests that individual needs might perhaps most effectively be met through the widest possible choice of programmes. But with few channels, minority interests cannot be met for more than a fraction of the time if majority tastes are to be catered for most of the time. And if the channels directly compete for viewers, minority interests will hardly be served at all: each channel will compete for the majority audience through similar offerings, as tends to occur on the national networks in the US.

A common argument is therefore that given a limited number of channels, viewer satisfaction should be maximised by complementary programming (e.g. Steiner, 1963). At any given time, the channels would be showing programmes which appeal to widely differing interest groups. (The programme choice available could also grow with any increased spread of cable television – potentially yielding vastly more channels — or technical developments like video-cassettes.)

The basic question then is what actual *use* the public would make of the greater variety of choices that might be on offer. Would complementary programming, with some items catering for specialist groups, be regarded as failing if students of literature still relax with "The Mary Tyler Moore Show" rather than watch "The White Devil" on the other channel, i.e. if even the minority audiences were largely to evaporate? This has been shown to occur to quite an extent by BBC2 in the UK and the public service channels in the US. "Minority" programmes are often hardly watched even by the relevant minority, if entertainment programmes are available simultaneously. This does not mean that no programme with obvious minority appeal should be screened, but that any programme decisions should be made whilst knowing what size and kind of viewing audience to expect.

128

A related problem concerns the broad programming policies of authorities like the IBA or BBC. The objective cannot be to *maximise* audience size in general. Individual programme companies and programme planners may already be trying to do that, and in any case it is not regarded as socially proper always to try for maximum audiences. But obviously the target cannot be to *minimise* audiences.

So an alternative target is set — *balance*. But we are now learning more of what balance in programming might actually mean *to the viewer*. It does not necessarily mean limiting the number of westerns that may be shown in any one week, implying that these would merely tend to cater to some groups of western addicts. There appear to be no such addicts. People who watch one particular western are no more likely to watch other westerns than are other viewers.

If an additional western were screened by ITV and achieved a rating of 20, then about 34% (i.e. 1.7×20) of the audience of any other ITV programme (whether another western or anything else) would watch the new western, and about 18% (i.e. $.9 \times 20$) of the audience of any BBC1 programme would so do. The same numbers would watch any other ITV programme with a rating of 20. So if "balance" means balancing programmes which *differ* from each other, we now know that in terms of aggregate patterns of programme choice, people do not differentiate between types of programmes.

Balance is also a widely used concept for controversial matters — political or other. But "putting the other side" in another broadcast might not seem very effective if few of those who were exposed to the first broadcast see the second (or vice versa). Yet this is what generally happens — about 55% if it is exactly a week later, and generally less if it is on another day of the same week. (The main exceptions are election and party political broadcasts in the UK, which are generally shown simultaneously on all channels.* They therefore achieve relatively high total ratings and correspondingly high duplication — heavy viewers do not switch off — although rather low appreciation scores.)

The conclusion for handling controversial matters might therefore be always to put both sides in the same broadcast. But that would often be retrogressive in terms of creative broadcasting, and possibly insulting to the audience. People are not always that easily influenced or manipulated

---

*Probably the first party political broadcast in the UK shown at *different* times occurred on 5 March 1975 (BBC1 and 2 at 9 pm, ITV at 10 pm), but at the time of writing, audience behaviour and appreciation have not yet been analysed.

by a single broadcast. (The evidence for election broadcasts is that they neither preach to the converted nor do the opposite – thus a Labour Party broadcast is no more likely to be seen by viewers of other Labour broadcasts than by viewers of Conservative or Liberal Party broadcasts.)

One of the simplest lessons of this book concerns people's low exposure to complete series of broadcasts. Producers, critics and the public need to learn that very few viewers of any series see all or most of its episodes. A series of programmes may form a unity to the producer but not to the audience. It is a mistake to suppose that on television anyone ever "reads the whole book through". One may doubt whether Marshall McLuhan knew it this way, but television certainly is not a "linear" medium.

The pull of the box

Such a finding may be difficult to square with the supposedly compulsive attraction of television. Surely a highly popular programme series will tend to be watched very regularly? It is true that *some* people will do so, but they will be a small minority. The popularity or high rating of the series largely arises from all the other, mostly irregular, viewers whom it attracts.

And even the "regular viewers" are often not all that regular. People may claim that they always watch their favourite programme. But they claim this only for one or two programmes and not for all the other twenty or so they see each week. And however regularly one thinks one watches a programme, one actually only watches it in those weeks when ones does not happen to be away or doing something else. (Like the football addict who watches "Match of the Day" obsessively about once every three weeks.)

Television is a medium where one hardly misses — is hardly even aware of — what one does not see. The pull of the box certainly exists when the set is on. But it seems to snap if one leaves the room and shuts the door, even in the middle of a programme. Suddenly the plot and the characters no longer matter so much. If one misses a favourite programme, how much does one really mind afterwards? How much does missing it reduce one's enjoyment when watching the next episode a week later? If one comes in half way through, how much effort does one really make to be filled in on what one has missed, or to see a repeat? The general run of programmes does not seem to be regarded as "serious" in that sense.

Even for "specialist" programmes (usually defined as such only if they have small ratings) the position is similar. They certainly do not attract more regular viewers. If anything they are viewed more by heavy (and hence by definition, relatively indiscriminate or catholic) viewers, and not so much by specialist or otherwise selective viewers. (Light viewers tend more to watch the popular programmes – that is why they are popular.)

The explanation seems to be that people with a real specialist interest do not generally feel a need to follow it *on television*. Artists do not feel they need to watch art programmes; knitters, knitting programmes; or businessmen, business programmes (unless perhaps some friend – or enemy – happens to be performing). Specialists already know all that. Even religious people do not religiously watch all their programmes, but go to church instead.

The physical attraction of the live screen does not mean that the programme material as such always exercises any great pull. (Repeat-viewing is not high and viewers of a certain programme do not necessarily watch others of the same type – the duplication of viewing law). A viewer develops a habit and hence a certain liking for a particular programme (e.g. "Ironside") and tends to watch it fairly regularly. But that does not mean that he feels the need to watch other almost identical-seeming programmes (e.g. "Cannon"). And most viewers also watch many programmes other than their favourites or the ones they "really like to watch", as we saw in Chapters 8 and 9.

The effects of television

Given the large amount of time many people spend watching television, it is not surprising that concern has been expressed about its effects, i.e. its possible influence on opinions and behaviour. Politicians, social observers, programme makers, and advertisers have all been conscious – perhaps too conscious – of the alleged power of television as a medium. As stressed earlier, we ourselves have little specific expertise in this wider area of assessment. But some brief comments seem necessary here.

Considerable social research into the effects of television has taken place in the UK, the US and various other countries. Sociologists and others have examined the impact of heavy television viewing on children (e.g. Himmelweit, Oppenheim and Vince, 1958; Belson, 1967; Halloran, Brown and Chaney, 1970; US Department of Health, Education and Welfare, 1972) and the role of television in politics (e.g. Klapper, 1960; Blumler and McQuail, 1968; Halloran, 1970). The relation of television to such topics as violence, education, the arts, and religion has also been studied (e.g. Halloran, 1970; ITA, 1970; Rubenstein et al., 1972; US Department of Health, Education and Welfare, 1972; Liebert et al., 1973;

131

Howitt and Cumberbatch, 1975). In addition, a vast and diffuse body of literature has been generated by journalists, politicians and other writers (some Further Readings are listed at the end of this book), but little firm clarification seems to have resulted.

A typical issue that has been widely discussed is the possibly stimulating effect of violence on television. Here it has now been widely recognised that people do not rush out into the street to imitate what they have just seen on the screen. Any effects must be more diffused. There is a good deal of violence around and it is only in that broader context that any additional effects of television can be understood.

Yet it is certainly true that westerns and other fictional television programmes continually portray a world in which both good guys and bad guys use violence to solve problems and achieve goals. One might feel that the effects can hardly be wholesome. But it has also been argued contrariwise that such formalised violence on television might be cathartic — reducing rather than increasing tension in the viewer. There is no simple answer.

Television has of course also brought *real* violence into the home. We now almost daily "see it how it is". Apart then from the medium possibly acting as a stimulus to further violence (or to demonstrations which can lead to violence), it can also be argued that overexposure may have anaesthetised many of us. We may have grown used to seeing violence and its effects (or hunger and *its* effects). But would we actually act any differently day by day if we had *not* grown callous from seeing violence? Passing by on the other side of the street is at least as old as the Bible. So what we have come to take for granted is perhaps only violence *on the screen*, and we need have become no more callous to the real thing than we already were. Indeed, as nations we may be becoming *more* concerned and socially conscious, perhaps because we are more informed and more aware. Typically, questions about the effects of television here are complex, and clear-cut answers largely non-existent.

A different charge is that of trivialisation. With the advent of television, cinema attendances and radio listeners declined. It has also been indicated that theatre going and the reading of books decreased among heavy viewers (Belson, 1966). But how much of a loss was that? How "cultured" was Broadway or the West End theatre of London, and how intellectual was the reading that was reduced? And what of the many who would not or did not go to the theatre or did not read? Do many of them now not occasionally see drama, or serialisations of classics (even if sometimes only because they do not switch off)? Have drama groups and local theatres really declined? And how do we allow for (and explain) the possibly greater financial support which society (government, business, the public) give to the arts? Assessment and evaluation is again difficult.

Another area is television advertising. Here the supposed power of television is more direct. The purpose is to influence specific attitudes and behaviour. Whatever the complexities of the advertising process may be, at least the aims and the means are relatively clear and explicit. Furthermore, there is a great deal of information directly relating to its expected effects — people's attitudes towards the advertised goods and their buying or consumption behaviour are widely measured. It is therefore possible to examine how advertising works through the media, as has been done more fully elsewhere (e.g. Ehrenberg, 1974). Here we briefly summarise the main arguments. This also provides lessons on the power of television more generally.

We note that advertising, like television itself, is in an odd position. Its extreme protagonists claim it has extraordinary powers and its severest critics believe them. Whilst television advertising can be effective, it is not as powerful as is often thought. There is no evidence that it works by any strong form of persuasion or manipulation.

The possible effects of advertising on the demand for whole classes of goods or services (e.g. cars, cigarettes) must be distinguished from its effects on people's choice between competitive makes or brands of the product (e.g. Ford, Fiat or Volkswagen). Many of advertising's critics believe it has powers to create consumer demand for goods and to build our acquisitive society. But product-class advertising as a whole — "Buy more cars", "Drink more tea", etc. — cannot be held responsible. For one thing, there is relatively little of this form of advertising; for another, it generally produces only minor results. There are no dramatic claims in the literature (if we have missed one, that is the exception).

Repetitive advertising for individual brands — "Buy Fords", "Drink Tetley's Tea", etc. — is where the bulk of television advertising is concentrated. This could lead to a higher level of consumption of the product-class as a whole than would exist without it, but there is no evidence that such secondary or even unintended effects (of *brand* advertising on total *product-class* demand) are either big or particularly common.

Whilst the bulk of advertising on television is for competitive brands, sales of these mostly do not actually change greatly from one year to the next. The great mass of brand advertising must therefore at best be *defensive* — the manufacturer aims to keep what he has, by helping to reinforce an already-existing consumer habit of buying his brand. Any feeling of satisfaction with the brand — that it is liked at least no less than

133

others available — has to be nurtured (with attitudes mostly changing *after* usage, so as to reduce feelings of cognitive dissonance). People mostly ignore advertising for brands or products which they are not already using — it says little to them — by a process of selective perception.

But occasionally new customers for a brand or product are created, or altogether new brands or products are launched. Here advertising can both build awareness and help lead to an initial trial. The ultimate test is whether the consumer likes it after he has had it; only then will a repeat-buying habit (or favourable word-of-mouth recommendation) develop. The most widely-quoted statistic is that nine out of ten new brands *fail*.

Advertising's role is seldom a very powerful one. It is not a matter of persuading or manipulating the ignorant consumer, since consumers of heavily advertised products are mostly highly experienced. They have usually already bought the product often and have used a wide range of different brands (Ehrenberg 1972). No exceptional liking or "image" needs to be induced in the consumer, because he knows similar brands to be similar and does not greatly care which he buys (which mainly matters to the *manufacturer*).

In its supposed role of creating our acquisitive society, advertising tends to be confused with the influence of the mass media generally. Consumers' awareness and expectations have been raised by magazines, films and television, and also by people's greater mobility and education, not merely by advertising as such. Advertising itself generally *follows* changes in habits and fashion, rather than leads. Many people want things which are hardly advertised at all.

The way in which advertising appears to work must also tell us about *other* effects of television. Thus the American President's National Commission on the Causes and Prevention of Violence in the late sixties argued that given that so many millions continue to be spent on television advertising to influence human behaviour, television advertising must be very powerful and hence television more generally must also have strong effects.

But if the effects of advertising itself are in fact relatively weak — mainly to reinforce those who are already using the product and are already well informed and highly experienced, and with non-users of the product hardly noticing the advertising — then we can expect the effects of the *programme* material on television to be even more diffuse. After all, the effects people worry about (like stimulating violence in the viewer, destroying his moral values, or increasing his expectations as a consumer) are usually not even what the programme makers are trying (or are paid) to sell.

Occasionally a television programme (or an advertisement) will leave viewers aware of something new (like a new form of violence or new attitudes towards others, or a new or previously ignored brand). But by itself this new awareness does not usually actually make one *try* the new thing, or give money to charity or be kind to foreigners, let alone will it inculcate a *habit* of doing so. The crucial factor is whether one likes the new thing after one has tried it. Rather than our attitudes causing behaviour, it is often the case that our more salient attitudes are affected by prior changes in our behaviour.

Any effect of television will tend to be slow and diffuse, and difficult to isolate from the effects of other factors. The effects will take place in a wider context. Showing affluence or cruelty will mostly influence people who are ready to be influenced. There is more sex in real life than is ever shown on the screen, where it is most noticed by those looking for it.

We must not exaggerate what to expect from television, nor use inappropriate yardsticks and norms to evaluate it. Many readers of this book will be people who tend to feel guilty when watching television (or at least when admitting to it). But the mass of the population do not feel like that. For them, watching television is not necessarily regarded as a weak-minded substitute for doing something "better".

It is true that the fare which television provides is largely repetitive and often seems mindless. But when as young children we were read bedtime stories, we often asked for the same story (or even a certain favourite page) to be read over and over again and again. Less privileged children have no bedtime stories read to them at all.

Television provides information, entertainment, and a way to pass the time in one's home. Much of it may be fairy stories, but there is nothing new about that. Some television may be good, most of it is mediocre, and some bad. To understand it better and perhaps in some way to improve it, we must examine television from the point of the consumer. How do we, the viewers, use it? That is what this book has tried to illuminate. If in doing so a number of myths and shibboleths about television *viewing* have been destroyed, this should also engender some meekness when pronouncing on its *effects*.

Summary

Little is firmly understood about the social or individual effects of television. Even advertising, widely thought to be supremely effective, is seldom very powerful and is mostly defensive in its role.

Appendix A:
Audience Measurement

The measurement of audience ratings, i.e. the numbers in the audience, accounts for the bulk of research into television audiences. The term "research" (as in "market research") is however something of a misnomer here for what is mainly a routine monitoring operation.

The measurement procedures used in different countries are often similar. Here we concentrate on those used in the United Kingdom, where both the independent television interests and the BBC operate continuous measurement systems covering viewing on all channels each day. But in their technicalities the two systems differ about as much as any two procedures for measuring the same thing could. We start with the JICTAR panel system, which has provided the data for the analyses in Chapters 2 to 6 of this book.

Measurement procedures

The JICTAR panels

The ratings operation for independent television is sponsored by JICTAR, the Joint Industry Committee for Television Advertising Research. This is composed of the ITV programme companies, advertising agencies and advertisers. JICTAR attempts to determine the requirements of its various members, finances the measurement programmes, and awards a contract to an independent research company to carry out the specified work. From 1955 to 1968 this contract was held by Television Audience Measurement Ltd (TAM) and since then it has been held by Audits of Great Britain Ltd (AGB).

The statistical basis of the ITV measurement procedure is an Establishment Survey conducted annually in a sample of about 23,000 homes throughout the country. This provides details about the number and types of television sets in use and the potential audience. Interviewers obtain information for each of the fourteen ITV regions about how many television sets are owned or rented, whether they are black and white or colour, which transmitters can be received, the quality of the reception, and socio-economic charcteristics of the households with television sets, and the approximate amount of television normally viewed (the latter being used as a control measure in the subsequent stages of subsampling).

With this data as a base, a panel of households is recruited by stratified random sampling in each area. In total about 2600 reporting panel homes in the UK are used each week, with 100 in the North East Scotland ITV area, 350 in London, and so on.

An electronic meter is attached to the television set in each sample home. The meter records on heat-sensitive paper tape when the set is switched on and off and to which station it is tuned, for each minute of transmission time. This provides minute by minute "ratings" or estimates of the percentage of television sets switched on to each channel.

Individuals in each of the homes (about 8000 people in total) also fill in a special viewing diary each week showing their personal viewing (and that of young children and guests) quarter-hour by quarter-hour.

The meter tapes and diaries are speedily processed and tabulated and the detailed information is conveyed to interested parties in Weekly Television Reports within eight to ten days of the end of the week in question. Additional reports published thrice yearly provide much more detailed data on audience composition.

Subscribers to the JICTAR service can also buy special analyses or computer tapes of the viewing data when they want to explore particular topics in more depth. One of the most common types of special analysis deals with the coverage and frequency of viewing schedules of commercials.

BBC audience research

The BBC's Audience Research Department parallels the JICTAR work in measuring audiences for each programme on each channel day by day, but the technique differs radically.

BBC interviewers approach a quota sample of over 2000 people a day, either at their homes or in the street, and elicit information about their television viewing and radio listening on the previous day. Different samples are interviewed each day, adding up to more than half a million people a year. The interviewers try to get respondents to review the previous day chronologically, using a "recall aid" listing the previous day's programmes.

This produces ratings, i.e. measures of the audience size for each programme, either in the population as a whole or in demographic subgroups, as from the JICTAR panels. But this "one-day aided recall" procedure produces no information on audience flow across different days or different weeks. The main reason for the BBC's approach, apart from tradition, is that at little extra cost the same procedure provides ratings for the BBC's radio audiences.

Other measurement procedures

Several other large scale audience measurement procedures have been used

137

in the past. In the early days of commercial television in the UK, Pulse Ltd ran one-day aided-recall surveys similar to the BBC's, but on a commercial basis. Granada TV Network Ltd in 1959–60 and JICTAR (through Marplan Ltd) in 1962–4 ran more elaborate seven-day aided recall surveys. These measured each informant's viewing over the previous seven days, using daily programme titles as recall aids. The purpose was to provide more detailed "audience composition" data on the viewing of subgroups of the population.

Other countries

Audience measurements in other countries are often along similar lines to ones used in the UK. There seems to be surprisingly little new to be learned. The main method used in the more advanced countries is the meter-based panel operation, as used by TAM and AGB in the UK and initially pioneered by the A. C. Nielsen Company in the United States.

Since there are more than 200 local markets in the US, this panel method is supplemented by seven-day self-completion diary surveys (*not* aided recall) operated by Nielsen and by Arbitron (American Research Bureau) for such local markets. The analyses in Chapter 7 were based on the latter company's data. Where telephone ownership is high, as in the US, *telephone* coincidental surveys also tend to be used (asking what one is viewing "now"). But in Mexico for example, where literacy is relatively low and the cost of any continuous measurement system too high, the main method used over some years has been periodic *personal* coincidental interview surveys.

Accuracy

The JICTAR meter/diary panel procedures in Britain were vetted in 1960 by Sir Maurice Kendall, then Professor of Statistics at the London School of Economics, and there has been a tradition of academic or near-academic consultants. In the United States, the Nielsen panel procedures and other services were commented on in 1961 by a working party of the American Statistical Association (Madow et al., 1961). These services were also subsequently examined, and at times stridently criticised, in extended hearings before the Oren Harris Congressional Committee on Legislative Oversight (Eighty-eighth Congress, 1964–5). The result is that the routine fieldwork and analysis procedures in the States are nowadays audited by firms like Price Waterhouse, under the aegis of the US advertising industry's Broadcast Ratings Council.

But generally there has been little attempt by those operating the various measurement techniques to establish the precise nature of any differences in the results or to develop the necessary adjustments. Rather, the aim has been to promote one's own preferred method.

In 1961 JICTAR however sponsored the TAM Comparison Survey, a substantial experimental comparison of various panel and aided recall techniques. This also included the "coincidental" method, where a household is questioned at a preselected time as to its viewing then. A limited comparison of some *operational* JICTAR and BBC data has also been made (Ehrenberg and Twyman, 1967), which also refers to some American results comparing meter panels and one-week diaries.

The main conclusion about the variety of different measurement procedures is that they tend to give similar results. There *are* differences, such as certain tendencies for JICTAR's measurements to favour ITV and the BBC's to favour the BBC (often picked upon by the press), but these are mostly not large and often seem to stem from different definitions. For example, the BBC surveys include anyone, whether or not he has a television set in his home, measure his viewing anywhere, "viewing" being determined as having seen at least half of a programme. In contrast, the JICTAR ratings used in such comparisons relate to an average minute by minute set-on measure for that part of the population which can reliably receive ITV transmissions. These differences would have to be allowed for in any meaningful comparison of results.

Some criticisms

Possible criticisms of the JICTAR system include the relatively small size of the panels used in each area, the potential inaccuracies of the self-reporting viewing diaries, and that only the "presence" of the viewer is measured by the 15-minute ratings, and whether the *set* is on by the minute by minute ratings. The BBC system is open to doubts about the methods of sampling and of interviewing (but aggregate sample sizes are very large when results are averaged over successive days or weeks). Still, the size, speed and attention to detail in both systems are impressive and it generally seems to be felt that they are adequate for the purposes for which they are mainly used.

Indeed, some contrary criticisms question whether these measurement efforts are not unduly large and elaborate. It was estimated some years ago that the JICTAR system involved making some 500 million measurements per year and that rating reports cover well over 10,000 pages of detailed statistics annually. It might be thought that nothing could vary so much as to require that amount of documentation or, if it did, that nobody could fully use the resulting data. Highly elaborate data collection may in fact reflect uncertainty about how the data should be used.

Presence and attention

Mechanical meter methods are widely used in audience measurement, even though they measure only whether the set is on rather than the usually more relevant factor of who is watching (as would be measured by self-completion diaries or aided recall interviewing).

Early doubts about the accuracy of methods involving human beings were one reason for the use of meters. Another is that meters can measure the movement of audience levels (in terms of sets on) minute by minute. This matters when ratings are used for assessing the audience to short advertisements. In contrast, most measurement procedures involving people's viewing attempt nothing more precise than measurement in quarter-hour time-bands, with viewing for at least 8 minutes out of the 15 notionally qualifying one as a "viewer". (Coincidental interviewing can in theory measure audience size at more precise points in time, but there are technical problems and it would be exceedingly expensive for full scale routine use, day in day out.)

One query which used to overhang meter ratings is whether anyone is actually watching. But it has been established over the years that the percentage of switched-on sets that are unattended tends on the whole to be negligibly small. This may however not be so for certain short time intervals, such as during commercial breaks. But the measurement of individual viewing behaviour in quarter-hour time-bands also breaks down for short time-periods such as commercial breaks. It is well known that substantial numbers of viewers leave the room momentarily during a commercial break. Furthermore, even those who remain are not all attending fully. A number of special studies were carried out in the early sixties to show the magnitude of such factors (as reviewed by Nuttall, 1962; Ehrenberg and Twyman, 1967). The main results are illustrated in Table A.1. They indicate that on average 20% of those present when programmes are shown will be absent during an intervening commercial break ("not present"), with more women than men being absent (30% versus 10%). For women the "not present" percentage tends to be higher early in the evening than later (40% versus 20%), but otherwise these types of figures are fairly steady throughout the evening. As for the other end of the scale, on average only about 40% of the adults in the programme audience are

Table A.1

Typical behaviour of the adult audience of TV programmes during a commercial break

UK, early 1960's	Men	Women	Adults
	%	%	%
All Programme Viewers	100	100	100
Viewing only	50	30	40
Viewing and otherwise active	30	30	30
Present but not viewing	10	10	10
Not Present	10	30	20

directly viewing during a commercial break, with another 40% or so being present but not necessarily giving their attention to the television fully or at all.

A variety of different measurement techniques has given very consistent results here. Table A.2 compares the results for coincidental questioning at the time of the break, for short term recall analysed over three different lengths of recall periods, and for longer term recall (using suitable programme aids in first establishing viewing of *programmes*) from interviews the following morning. (See Ehrenberg and Twyman, 1967, for sources and further discussion.)

Table A.2

Programme viewers' behaviour during commercial breaks, as measured by different measurement techniques

UK, early 1960's	MEASUREMENT TECHNIQUE					
	Coinci-dental	Recall (in minutes)			Recall 12 hrs	Average (approx)
		0-19	20-39	40-60		
	%	%	%	%	%	%
All Programme Viewers	*	100	100	100	100	100
Viewing only	42	40	39	38	40	40
Viewing and active	31	31	30	31	33	30
Present, not viewing	7	9	8	8	6	10
Not present	*	20	21	21	17	20

* The definition of "not present" differs

141

It follows that factors of momentary absence from the viewing room or details of viewers' greater or lesser involvement in other activities can be reliably measured. The available evidence is that in the aggregate the patterns are fairly stable and often influenced by social habits rather than by the nature of the programme material as such. To what extent audience sizes should be routinely expressed in terms of "presence" or "attention" remains a moot point. But in many cases it does not appear to affect one's broad conclusions when comparing audience sizes at different points in time (see also Twyman, 1971, for a review).

Audience reactions

However defined and measured, audience size is by itself an inadequate or incomplete index of viewer appreciation of a particular programme or of relevant response to a commercial. Audience size may not indicate audience interest or appreciation because of inconvenient transmission times, the influence of other family members, and programme clashes on different channels. In addition people vary in the extent and in the manner in which they appreciate the programme they watch. Both the BBC and independent television interests have therefore studied audience response in various ways.

The IBA has encouraged a number of research efforts at measuring audience reactions on some continuing basis. The earliest was the TVQ system introduced by TAM in 1964, using postal questionnaires sent to a sample of viewers who rated the programmes on a 5-point scale from "One of my favourites" to "Poor". This system was criticised as being too simple and the IBA commissioned some new research into ways of assessing programmes (Frost, 1969). In parallel, Rothman and Rauta (1969) tried to create a greater understanding of the pattern of audience appreciation by creating a viewer typology, categorising people by the types of programmes they liked or disliked.

More recent work on monitoring audience appreciation, originally commissioned externally by the IBA and now run by it as an internal service, is on a more or less continuous "panel" basis. Panel members post in alternate weeks reports on programmes viewed on all three channels, using an overall appreciation scale for all programmes on the three channels, and individual comments. The work has been further developed under the name AURA which operates with the help of a representative panel of about 1000 adult viewers in London and random postal samples of about 2000 in other ITV areas. Some findings have been discussed in Chapter 8.

The BBC has also been explicit about the limitations of relying on audience size figures to judge programme performance. Its Audience Research Department uses a regular monitoring of audience reactions. A panel of 2000 members are recruited either by public appeal or by direct invitation to persons previously interviewed in the regular aided recall surveys. (The panel may therefore not be particularly representative.) The enrolled panel members are sent weekly batches of questionnaires about a wide variety of programmes. These vary according to the type of programme.

In simple cases the panel member is asked to rate the programme in different ways, e.g. to indicate for a comedy programme to what extent it was funny, vulgar, boring, etc. These rating scales provide data for Reaction Profiles that show in graphical form the extent to which the panel favourably described the programme. For example, a play may receive Reaction Profile indices of interesting/boring 80:20, well presented/poorly presented 70:30, etc.

Sometimes the BBC also uses longer questionnaires to provide material for programme reports which try to give a balanced picture of the opinions expressed, placing correct emphasis on majority and various minority views (BBC, 1974, 1975). More formal attempts have also been discussed to measure the "gratification" that viewers receive from different programmes (e.g. Emmett, 1966).

It is widely accepted that in assessing television we do not merely want to count heads (the ratings), but also "the number of heads which are contented, dissatisfied, or indifferent", as Brian Emmett of the BBC has phrased it. This is an area of audience measurement where more work is needed. Nonetheless, the more detailed cross-analysis of *ratings data* as in Chapters 2 to 7, also provides some insights into what programmes mean to people.

Summary

The bulk of audience research (e.g. by ITV interests and the BBC in the UK) is concerned with the routine monitoring of audience size — the ratings. Very different measurement procedures are used, but the resulting data mostly appear adequate for their purpose.

The extent to which people nominally measured as part of the audience are in fact always present or attending has also been studied. This has led to consistent findings.

Appendix B: Updating to 1974

It is fundamental to this book to know how widely the results discussed do in fact hold. Our main findings derive from audience measurement data in the UK up to 1971 and we felt that a check on more up-to-date data was essential. Analysis of data for 1974 was therefore put in hand, and the results are summarised in this Appendix.

The data analysed are viewing records from the AGB/JICTAR panel in London over a period of four weeks in May 1974 for a sample of 585 adults (aged 16 and over) in homes with TV sets able to receive ITV and BBC1 (over 90% could also receive BBC2).

Repeat-viewing

Repeat-viewing has been analysed for successive episodes of 24 ITV and 27 BBC programmes where the ratings showed little or no change from one week to the next. As with the earlier data discussed in Chapter 5, the average for repeat-viewing was again almost 55%. More precisely, the repeat-viewing levels were:

average ITV programme 53%,
average BBC programme 49%.

The slightly lower repeat-viewing average for BBC programmes seems largely due to their somewhat lower rating levels, but by the same token the precise degree of agreement with the previous repeat levels may be more apparent than real, since the programmes analysed included more low rating programmes than before.

There are several other findings for the 1974 data. Firstly, the data confirm the earlier tendency for the repeat-viewing level to decrease with the rating level. If anything, the trend is stronger than that indicated by the results in Table 5.3, as had also been noted in subsequent work referred to in Chapter 5. It is therefore becoming increasingly clear that the repeat-viewing level is not a constant, i.e. generally 55% with only irregular variation above and below. Nonetheless, the figure of 55% continues to provide an effective, if oversimplified, summary of the main finding.

Secondly, there is again no "erosion" of repeat-viewing for weeks further apart (cf. Table 5.8). Thus over a four-week period we have

% of 1st week audience who viewed in:	1971	1974
second week	54	53
third week	51	53
fourth week	51	51

Thirdly, repeat-viewing at weekends is lower than on weekday evenings (cf. Table 5.5):

% repeat-viewers of	1971	1974
weekend programmes	49	40
weekday programmes	57	59

The larger difference in 1974 is in line with a relatively lower average rating at the weekend.

Fourthly, there are once more almost no generalisable differences in repeat-viewing levels for different types of programmes. Films were rather low in 1974 but not in 1971 or 1969 (Tables 5.5 and 5.7). Serials are high, but there were few in the 1974 data analysed, and these consisted primarily of "Coronation Street" and "Crossroads", each a long-running serial with two or more episodes a week. Such serials have generally shown relatively high week-by-week repeat-viewing levels (cf. Table 5.7 for 1969 and the 63% for "Coronation Street" in 1971 in Table 5.3). However, these variations in the repeat-viewing levels for films, serials, etc. all seem to reflect the more general relationship between repeat-viewing and rating.

Audience duplication

The duplication of viewing law holds again in 1974. The duplication coefficients for pairs of weekday programmes are very much the same as those for 1971 (Table 3.11):

Within-channel	1971	1974
ITV × ITV	1·7	1·8
BBC1 × BBC1	1·8	1·8
BBC2 × BBC2	1·9	2·2
Between-channel		
ITV × BBC1	0·8	0·9
ITV × BBC2	0·9	0·9
BBC1 × BBC2	1·1	1·1

The proportion of viewers of an ITV programme who watched another programme on another day was therefore about 1·8 times as high as the latter's rating if it was another ITV programme, and only 0·9 times as high if it was a BBC programme.

The usual exceptions also recur, i.e. higher duplications for pairs of late afternoon/early evening and for pairs of late evening programmes, and audience duplications involving one or more week-end programmes are again fractionally lower, as was also the case previously (see Table 4.5).

The deviations from the duplication law are again mostly small, an average of ·8 rating points for 302 pairs of ITV programmes, ·5 for 224 BBC1 programme pairs, and ·6 for 586 cross-channel programme pairs (ITV and BBC1). Systematic sub-pattern for programmes of a particular type can therefore exist only within these relatively narrow average limits. But as before, (e.g. in Table 4.10 for 1967), there is no evidence of any dramatic deviations for any of the conventional (or commonsense) programme classifications that have been checked. Thus for week-day programmes (excluding the early-evening or late-evening pairs), the average deviations from the duplication of viewing law by type of programme are (in rating points):

	1967	1974
News	−·3	·1
Current affairs and comment	·2	−·1
Plays	·7	·0
Adventure and crime series	·1	·5
Comedy series	·1	·3
Other light entertainment	·3	·7

In the week-day evening programmes analysed so far there was insufficient data to consider sports programmes or films.

The average ratings of the programmes are about 15 to 20, implying an average duplicated audience within-channel of about $1·8 \times 20 \times 20/100 = 7\%$ of the population for pairs of programmes on the same channel, and about 4% for pairs on different channels. Against this, the above deviations are still small, but evidence is possibly building up for a small cluster of light entertainment programmes.

The inheritance effect

Audience inheritance for programmes on the same channel again occurred in 1974. A programme shared a higher proportion of its audience with another programme if this was shown earlier that evening rather than on a different day.

Previously, this inheritance effect had usually lasted only for the adjacent and, to a lesser extent, the adjacent-but-one programme (both in the UK and in the US). But in the UK in May 1974 it appears that the effect often lasted over a larger number of programmes. Thus many programme pairs on a given evening which had several other programmes intervening would have more viewers in common than predicted by the between-day duplication of viewing law. This longer-lasting inheritance effect is closer to what many people in television had in the past thought occurred generally, but had in fact never been observed. Neither the extent nor the duration of the effect must be exaggerated, but more analysis is now needed to describe and understand this new phenomenon.

There are several possible reasons for such a change in viewing behaviour (if it is confirmed as holding generally). For example, there seems to have been an increase in the extent of non-coterminous programming (BBC programmes now seldom start on the half-hour or hour, as ITV ones tend to). Again, there has been increased on-air promotion of subsequent programmes (especially by the BBC). Thirdly, there is now a marked tendency for relatively long programmes, lasting about 1½ hours, fairly early in the evening, e.g. from 7.30 to 9.00 p.m. If such a programme is followed by another hour-long programme on one of the channels, it is likely that channel-switching will be inhibited.

Other work

Further analysis of the 1974 data will in due course lead to a good deal of new knowledge. It is our first data tape which includes results for children's viewing and which also covers data for more than four weeks. Children's viewing habits and viewing patterns over longer periods of time will therefore be among the new areas of research.

Summary

Repeat-viewing and audience duplication in 1974 were very much as in previous years. The inheritance of audiences from preceding programmes on the same day appears however to be more extensive than in the past.

IBA Reports

Qualitative Impressions of Audience Reach	Sept.	1967
The Factor-Analytic Search for Programme Types	Oct.	1967
Duplication of Viewing Between and Within Channels	Jan.	1968
The News in May	Feb.	1968
The Standard Programme Categories	April	1968
The Limits of the Inheritance Effect	June	1968
The News at Ten	June	1968
The Size of the Inheritance Effect	Aug.	1969
Higher Duplication	Aug.	1969
Sex and Age in Duplicated Viewing	Sept.	1969
Repeat-Viewing Week-by-Week	March	1970
TV Duplication of Viewing in June 1969	April	1970
Repeat-Viewing with Irregular Programming	Aug.	1970
Party Political Broadcasts	Nov.	1970
World Cup '70	March	1971
Housewives Viewing Intensity	Sept.	1971
Twenty Questions and Answers about Channel-Loyalty	Jan.	1972
The Audience of the News	Jan.	1972
Audience Build-up — Some Preliminary Findings	April	1972
Viewing Intensity and Programme Choice : Total TV	Dec.	1972
Viewing Intensity and Programme Choice : ITV	Dec.	1972
Repeat-Viewing and Audience Cumulation	Jan.	1973
Repeat-Viewing and the Audience Appreciation Index	April	1973
Audience Reaction to Different Episodes	July	1973
Viewing Intensity and Programme Choice (Supplement)	July	1973
Duplication of Viewing Amongst Regular Viewers	Aug.	1973
Mr Trimble — The Viewing of a Pre-School Programme	Nov.	1973
Repeat-Viewing by Light Viewers — An Initial Analysis	Feb.	1974
Duplication of Viewing in One-Person Households	March	1974
Duplication of Viewing among Heavy and Light Viewers	April	1974
Set-On Data	Aug.	1974
Programme-Type Effects	Oct.	1974
55% Repeat-Viewing	Nov.	1974
Audience Retention Within a Programme	Nov.	1974
The Liking of Programme-Types	Jan.	1975
Repeat-Viewing among Light, Medium and Heavy Viewers	Jan.	1975

Further Readings

Audits of Great Britain, *A Review of Television Audiences, July 1968 – July 1971*, London, AGB, 1971.

Arons, L. and May, N.A. (eds), *Television and Human Behavior*, New York, Appleton, Century, Crofts, 1963.

Belson, W.A., *The Impact of Television*, London, Crosby Lockwood, 1967.

Bogart, L., *The Age of Television*, New York, Frederick Ungar, 1972.

Bower, R.T., *Television and the Public*, New York, Holt, Rinehart and Winston, 1974.

British Broadcasting Corporation, *Annual Review of BBC Audience Research Findings*, London, BBC, 1974.

British Broadcasting Corporation, *BBC Handbook 1975*, London, BBC, 1974 (or subsequent years' editions).

Broadbent, S.R., *Spending Advertising Money*, London, Business Books, 1970.

Ehrenberg, A.S.C., 'Repetitive Advertising and the Consumer', *Journal of Advertising Research*, 14, No. 2, pp. 25–34, 1974.

Halloran, J.D., *The Effects of Mass Communication*, Leicester University Press, 1964.

Halloran, J.D., *The Effects of Television*, London, Panther Books, 1970.

Himmelweit, H.T., Oppenheim, A.N., and Vince, P., *Television and the Child*, London, Oxford University Press, 1958.

Independent Broadcasting Authority, *IBA 1975: A Guide to Independent Television*, London, IBA, 1974 (or subsequent years' editions).

Klapper, J.T., *The Effects of Mass Communication*, New York, Free Press, 1960.

Liebert, R.M., Neale, J.M. and Davidson, E.S., *The Early Window*, Oxford, Pergamon Press, 1975.

Mayer, M., *How Good are Television Ratings?*, New York, Television Information Office, 1966.

Rubenstein, E.A., Comstock, G.A., and Murray, J.P. (eds), *TV and Social Behavior*, Washington DC, Government Printing Office, 1972.

Steiner, G.A., *The People Look at Television*, New York, Alfred A. Knopf, 1963.

References

BBC (1970), *BBC Audience Research in the United Kingdom: Methods and Services,* London, BBC.

BBC (1972), *Violence on Television, Programme Content and Viewer Perception,* London, BBC.

BBC (1974), *Annual Review of BBC Audience Research Findings, No. 1,* London, BBC.

Barnett, T., and Lougher, J. (1971), 'Multiplus – a model for television planning and evaluation', *Admap,* January, pp. 20–5.

Belson, W.A. (1967), *The Impact of Television,* London, Crosby-Lockwood.

Blumler, J.G., and McQuail, D. (1968), *Television in Politics: Its Uses and Influence,* London, Faber.

Bower, R.T. (1973), *Television and the Public,* New York, Holt, Rinehart and Winston.

Bruno, A.V. (1973), 'The network factor in TV viewing', *Journal of Advertising Research,* vol. 13, no. 5, pp. 33–9.

Ehrenberg, A.S.C. (1966), 'Laws in marketing – a tailpiece', *Applied Statistics,* vol. 15, pp.257–67 (also in Ehrenberg, A.S.C. and Pyatt, F.G. (eds, 1971), *Consumer Behaviour,* London, Penguin Books.)

Ehrenberg, A.S.C. (1968), 'The factor analytic search for programme types', *Journal of Advertising Research,* vol. 8, no. 1, pp. 55–63.

Ehrenberg, A.S.C. (1972), *Repeat-Buying,* Amsterdam, North Holland and New York, American Elsevier.

Ehrenberg, A.S.C. (1974), 'Repetitive advertising and the consumer'. *Journal of Advertising Research,* vol. 14, no. 2, pp. 25–34.

Ehrenberg, A.S.C. (1975), *Data Reduction,* London and New York, John Wiley.

Ehrenberg, A.S.C., and Goodhardt, G.J. (1969a), 'Duplication of viewing between and within channels', *Journal of Marketing Research,* vol. 6, pp. 169–78.

Ehrenberg, A.S.C., and Goodhardt, G.J. (1969b), 'Practical applications of the duplication of viewing law', *Journal of the Market Research Society,* vol. 11, pp. 6–24.

Ehrenberg, A.S.C., Goodhardt, G.J. and Haldane, I.R. (1970). 'The news in May', *Public Opinion Quarterly,* vol. 33, pp. 546–55.

Ehrenberg, A.S.C., and Twyman, W.A. (1967), 'On measuring television audiences', *Journal of the Royal Statistical Society, A.,* vol. 130, pp. 1–59.

Eighty-Eighth Congress (1964–5), *Hearings before the Sub-committee on Legislative Oversight of the Committee on Inter-Estate and Foreign Commerce on Methodology, Accuracy and Use of Ratings in Broadcasting,* Washington DC, Government Printing Office.

Emmett, B.P. (1966), 'The design of investigations into the effects of radio and television programmes and other mass communications', *Journal of the Royal Statistical Society A.,* vol. 129, pp. 26–60.

Fawley, A., and Fairclough, E. (1972), 'The prediction of coverage and four plus for television schedules', *Admap,* February, pp. 74–6.

Frost, W.A.K. (1969), 'The development of a technique for TV programme assessment', *Journal of the Market Research Society,* vol. 11, pp. 25–44.

Gensch, D. and Ranganathan, B. (1974), 'Evaluation of television program content for the purpose of promotional segmentation', *Journal of Marketing Research,* 11, pp. 390–8.

Goodhardt, G.J. (1966), 'The constant in duplicated television viewing', *Nature,* vol. 212, no. 5070, p. 1616.

Goodhardt, G.J., and Chatfield, C. (1973), 'Gamma distribution in consumer purchasing', *Nature,* vol. 244, no. 5414, p. 316.

Halloran, J.D. (ed., 1970), *The Effects of Television,* London, Panther Books.

Halloran, J.D., Brown, R.L., and Chaney, D.C. (1970), *Television and Delinquency,* Leicester University Press.

Himmelweit, H.T., Oppenheim, A.N., and Vince, P. (1958), *Television and the Child,* London, Oxford University Press.

Howitt, D. and Cumberbatch, G. (1975), *Mass Media Violence.* London, Paul Elek.

Hulks, R., and Thomas, S.G. (1973), 'PREFACE: A simple model for the prediction of television coverage and frequency distribution', *Admap,* December, pp. 349–53.

Hyett, G.P. (1958), *The Measurement of Readership,* Statistics seminar at the London School of Economics.

ITA (1970), *Religion in Britain and Northern Ireland: A Survey of Popular Attitudes,* London, ITA.

Johnson, D., and Peate, J. (1966), 'The estimation of television viewing frequency', *Admap,* July/August, pp. 394–6.

Klapper, J.T. (1960), *The Effects of Mass Communication,* New York, Free Press.

Kirsch, A.D., and Banks, S. (1962), 'Program types refined by factor analysis', *Journal of Advertising Research*, vol. 2, no. 3, pp. 29–32.

Liebert, R.M., Neale, J.M. and Davidson, E.S. (1973), *The Early Window*, Oxford and New York, Pergamon.

Madow, W.G., Hyman, H.H., and Jessen, R.J. (1961), *Evaluation of Statistical Methods Used in Obtaining Broadcast Ratings*, Washington DC, Government Printing Office.

Marplan (1965), *Report on a Study of Television and the Managerial and Professional Classes*, London, Marplan.

McPhee, W.N. (1963), *Formal Theories of Mass Behavior*, New York, Free Press.

Metheringham, R. (1964), 'Measuring the net cumulative coverage of a print campaign', *Journal of Advertising Research*, vol. 4, pp. 23–8.

Nuttall, C.G.F. (1962), 'TV Commercial Audiences in the United Kingdom', *Journal of Advertising Research*, vol. 2, no. 3, pp. 19–28.

Pilkington Committee (1960), *Report of the Committee on Broadcasting*, London, HMSO.

Rothman, L.J., and Rauta, I. (1969), 'Towards a typology of the television audience', *Journal of the Market Research Society*, vol. 11, pp. 45–69.

Rubenstein, E.A., Comstock, G.A. and Murray, J.P. (eds, 1972), *TV and Social Behavior*, Washington DC, Government Printing Office.

Schuchman, A. (1968), 'Are there laws of consumer behavior', *Journal of Advertising Research*, vol. 8, pp. 19–27.

Segnit, S. and Broadbent, S. (1973), 'Life style research', *European Research*, 1, pp. 6–13.

Steiner, G.A. (1963), *The People Look at Television*, New York, Alfred A. Knopf.

Swanson, C.E. (1967), 'The frequency structure of television and magazines', *Journal of Advertising Research*, vol. 7, no. 2, pp. 8–14.

Twyman, W.A. (1971), *A Review of Research into Viewership of Television Commercial Breaks*, London, Institute of Practitioners in Advertising.

US Department of Health, Education and Welfare (1972), *Television and Growing Up: The Impact of Televised Violence*, Washington DC, Government Printing Office.

Williams, R. (1974), *Television*, London, Fontana.

Index

Indexer's note: specific television programmes referred to in the text are listed together under 'Programme titles'.

AB social classes 127
Adventure programmes/series 109, 116, 121f, 146
Advertisements/advertising 1–4, 6, 66, 70, 98, 133–5 (effects)
Afternoon viewing see Off-peak viewing
Aided-recall procedures 43, 137–9, 141
American Broadcasting Company (ABC) 2, 15, 19, 85–7, 89, 92
American Research Bureau (Arbitron) xi, 7, 85, 87, 138
American Statistical Association 138
Appreciation Index 8, Chapter 8, 114
Arons, L. 149
Arts and TV 131
Arts documentaries/programmes 14f, 131
Aske Research Ltd. vii, xi
Attitudinal measures 8f
Audience:
    appeal 17, 115f
    appreciation xi, 6, 8, Chapter 8, 125, 127, 142f
    appreciation panels 8
    characterisation/make-up 9, 78, 138
    cumulation 64–73
    duplication x, 9–15, 25f, 37, 55, 75, 79, 84–95 (US), 125f, 145–7:
        BBC1 and BBC2 34f
        BBC1 and ITV 29f, 35
        BBC2 and ITV 34f
        on non-consecutive days 28
        of same programme see Repeat-viewing
    flow x, 7, Chapter 2, 23 (across channels, 60, Chapter 7 (US), 137
    measurement vi, 6–8, Appendix A
    overlap see Audience duplication
    polarisation 39
    presence 140–2
    reactions 142f
    size see Ratings

Audits of Great Britain (AGB) vii, xi, 7, 67, 136, 138, 144, 149
AURA 142
Availability factor 13, 33, 54, 59, 62, 91, 126
Average frequency 66–8

Banks, S. 45, 151
Barnett, T. 67, 150
BBC1 2, 8, 76 (hours viewed)
BBC2 2, 8, 39, 41 (high duplication), 76 (hours viewed), 97, 128
Belson, W.A. 131f, 149, 150
Beta Binomial Distribution (BBD) Model 65–73
Birmingham, Alabana 85
Blumler, J.G. 131, 150
Bogart, L. 149
Bower, R.T. 127, 149, 150
British Broadcasting Corporation (BBC) 2, 6, 8, 43, 76, 82, 98, 129, 136f, 149, 150
BBC audience research 7, 137, 139, 143
Broadbent, S.R. 109, 149, 152
Broadcast Ratings Council 138
Brown, R.L. 131, 151
Bruno, A.V. 49, 150
Business programmes 131

Chaney, D.C. 131, 151
Channel loyalty vii, ix, x, 12, Chapter 3, 23 (defined), 37, 43, 45, 49, 89, 92, 95, 126:
    to BBC1 31f
    to BBC2 32–4
    to ITV 24–9
Channels:
    choice of 2f, 75–7
    number of ix, 2, 95
    public service 4, 128
Channel switching ix, 29–31, 89, 95, 106, 147

Chatfield, C.   67, 151
Children's programmes   5, 12, 122
Coincidental interview method   138—41
Collins, M.A.   vii
Columbia Broadcasting System (CBS)
   3, 49, 85—7, 92, 94
Commercial breaks   140—2
Community stations   3
Comstock, G.A.   149, 152
Congressional Committee on Legislative
   Oversight , 138
Crime programmes   4, 146
Cumberbatch, G.   132, 151
Current affairs programmes   5, 17, 105f,
   109—11, 116, 120—2, 146

Data analysed   xi, 144
Davidson, E.S.   149, 152
Demographic factors   37f, 58
Diary surveys   138f
Double jeopardy   59
Doyle, Peter   xi
Drama programmes   4f, 17, 58, 60, 132,
   146
Duplication coefficient (k)   11, 20, 24,
   30, 35f, 38, 82, 89, 92, 145f
Duplication of viewing law   x: 11—15
   (defined), 18—21, 26—30, Chapter 4,
   38f (time factor), 50f (news bulletins),
   79—83 (and viewing intensity), 85,
   89—95 (US), 126, 131, 145—7:
   deviation/exceptions   13, 18, 28, 31,
   39, 146
   empirical basis   13
   theoretical explanation   126

Early evening:
   effect   39f
   viewing see Off-peak viewing
Education and TV   131
Educational:
   networks   3
   programmes   4
Effects of television   ix, 131—5
Ehrenberg, A.S.C.   vii, 18, 49, 59, 85,
   133f, 139f, 149, 150f
Eighty-Eighth Congress   138, 151
Election broadcasts   129f
Emmett, B.P.   143, 151
"Enforced" viewing   120, 123
Establishment Survey   136
Exposure distributions   64—73

Factor analysis   49
Fairclough, E.   67, 151
Fawley, A.   67, 151
Federal Communications Commission
   (FCC)   3f
Films   5, 17, 60, 125, 145f
France   2
Frequency of viewing   102—5 (and AI),
   107, 122f, 126
Frost, W.A.K.   142, 151

Games and quiz programmes   5
General entertainment programmes
   see Light entertainment programmes
Gensch, D.H.   49, 151
Germany   2
Goodhardt, G.J.   vii, 67, 81, 150f
Government monopoly   2f
Granada Television Network Ltd.   138

Haldane, Ian R.   viii, xi, 150
Halloran, J.D.   131, 149, 151
Harris, Oren   138
Heavy viewers   xi, 75—82, 85 (US),
   104, 112
Himmelweit, H.T.   131, 149, 151
Howitt, D.   132, 151
Hulks, R.   67, 151
Hyett, G.P.   68, 151
Hyman, H.H.   152

Independent Broadcasting Authority
   (IBA)   vii, xi, 2, 8, 98, 112, 129, 142,
   149
Independent channels/stations   4, 6, 92,
   95
Independent Television Authority (ITA)
   131, 151, see also IBA
Inheritance effect   vii, 13, 37, 43—6, 52,
   85, 90, 92—5 (US), 97, 126, 147
Intellectual programmes   116
Intensity of viewing see Viewing intensity
Intentions to buy   98
Irregular viewing   122f, 125
ITV   6, 76 (hours viewed), 82, 98, 136

Jessen, R.J.   152
Johnson, D.   67, 151
Joint Industry Committee for Television
   Advertising Research (JICTAR)
   vii, xi, 7, 39, 67, 101, 136—9, 144

Kendall, Sir Maurice 138
Kirsch, A.D. 45, 151
Klapper, J.T. 131, 149, 151

Lapsed viewers 62
Las Vegas 85
Late-evening viewing *see* Off-peak viewing
Lead in effects 45f, 93f
Lead out effects 46, 93f
Leo Burnett Company xi, 109, 113
Leo Burnett Life Style Research Study
  109, 120, 122
Lewis, H.B. xi
Liebert, R.M. 131, 149, 152
Light entertainment programmes 5, 109,
  112, 114–22, 128, 146
Light viewers xi, 75–82, 85 (US), 104,
  131
"Liking" Chapter 9
Lougher, J. 67, 150

Madow, W.G. 138, 152
Marplan 127, 152
May, N.A. 149
Mayer, M. 149
McPhee, W.N. 59, 152
McQuail, D. 131, 150
Media planning 66
Meter panels 136–40
Metheringham, R. 68, 152
Minority interests/programmes 128–31
Murray, J.P. 149, 152
Musical programmes 5

National Association of Broadcasters 4
National Broadcasting Company (NBC)
  3, 85–7
National Commission on the Causes and
  Prevention of Violence 134
Neale, J.M. 149, 152
Networks 2, 6:
  national 3, 7, 23, 85, 89, 91, 95, 128;
  regional 4, 85, 95
News programmes 5f, 15–7, 49–54, 146
New York 15, 85, 89
Nielsen, A.C., Company 7, 138
Non-repeat viewing 63
Non-stationarity 59–61, 68–70
Nuttall, C.G.F. 140, 152

Off-peak viewers/viewing 37, 39–41,
  90f (US), 126, 146
One-person households 49

Opinion research 7
Oppenheim, A.N. 131, 149, 151

Peate, J. 67, 151
Persuasion 133
Pilkington Committee 97–9, 152
Political broadcasts 129f
Politics and TV 131
Population 18 (defined)
Postal questionnaires 142
Prediction 66
Presence of audience 140–2
Price Waterhouse 138
Programme:
  appeal *see* Audience appeal
  categories/types 4 (defined), 46–9,
    57f, 70, 105f, Chapter 9, 125, 145f
  choice (actual behaviour) x, 8, 75,
    77f, 81f
  clusters 106, Chapter 9, 127
  companies 2, 39
  content 4, 12, 15, 17, 46, 55, 71,
    125f, 142:
    effects 37, 46–9, *see also* Effects
      of TV
  loyalties ix
  material 134
  type preferences ix–xi, 46–9, 54,
    Chapter 9, 119–23 (*vs.* viewer
    behaviour)
Programmes:
  available 4–6, 128
  "cult" 113
  high-rated 82, 99
  long 147
  low-rated 82, 99
  thirty-minute 19
Programme titles:
  ABC News 15, 19
  Adventurers, The 24–26, 40
  Aquarius 116
  Bless This House 26, 40
  Boxing 114
  Brady's Bunch 89
  Cannon 131
  Cinema 14
  Contours of Genius 14
  Coronation Street 60, 69f, 104f, 116,
    145
  Crossroads 145
  Eyewitness News 15
  Family at War 155f
  Flintstones, The 6

4.30 Movie, The   15
Galloping Gourmet, The   116
Golden Shot, The   114
Hawaii 5—0   109
Ironside   131
Jason King   116
Lassie   6
Line-Up   116
Lost in Space   39
Love American Style   15
Lucy Show, The   6, 86
Man at the Top   101, 103
Mary Tyler Moore Show, The   128
Match of the Day   114, 130
Midweek   17
Mind of J.G. Reeder, The   24f, 46
Mr Trimble   64f
Monday Play, The   17
Monty Python   116
Morecambe and Wise   114
News at Ten   10, 25, 40, 46, 53f
Opportunity Knocks   25, 114
Owen, M.D.   116
Panorama   16, 105f, 116, 120, 122
Persuaders, The   109
Peyton Place   116
Public Eye   116
Rugby Special   110, 114
Saint, The   26, 40, 46
$ Six Million Man, The   89
Startrek   6, 16
Talk-Back   116
This is Your Life   28, 65f, 68, 116
This Week   105
This Week — The Arts   14
Today   39
Tuesday Film, The   17
World in Action   25, 46, 105f
World of Sport   110, 114
Wrestling   114f
Z-Cars   55f, 59
Programming   ix, Chapter 1:
   balance/variety   2—4, 8, 129
   competitive   8, 34, 54, 97
   complementary   2 (defined), 8, 34,
      53f, 128
   coterminous   89
   irregular   61f, 125
   non-coterminous   2 (defined), 147
   policy   2, 37, 49—54, 128f
   repetitive   5f, 16
Promotion, on-air   147
Proportionality factor see

Duplication coefficient
Public affairs programmes see Current
   affairs programmes
Pulse Ltd.   138

Quiz programmes see Games and quiz
   programmes

Ranganathan, B.   49, 151
Ratings   xf, 6f, 11, 13, 16f, 18 (defined),
   26, 55, 77, 88f, 97, 99 (and AI), 104f,
   107, 122 (re viewing claims), 125—7,
   136—42, 144—6; and repeat-viewing
   56f, 59—61, 79, 88, 144f
Rauta, I.   142, 152
Reach   65—8
Reaction profiles   143
Recall measurement system (BBC)   43,
   137
Religion and TV   131
Religious programmes   131
Repeat-viewing   x, 6, 9, 15—7 (defined),
   Chapter 5, 63 (non-consecutive
   episodes), 75, 78f, 85—9 (same week),
   95, 97, 100—7 (and AI), 125, 128,
   131, 144f, 147:
   independent stations, networks   87f
   individuals   101—5
   national networks   86f
Re-runs   86
Rothman, L.J.   142, 152
Rubenstein, E.A.   131, 149—52

Sampling errors   17, 89
Schedules, mixed   68, 70f, 73
Scheduling   55, 62
Schuchman, A.   59, 152
Seasonal factors   42
Segnit, S.   109, 152
Selective viewing   77—9, 82, 131
Serials   4f (defined), 17, 60, 125, 145
Series   4f (defined), 17, 55, 60, 126, 128
Situation comedy programmes   4, 46
Social research   131
Sports programmes   5, 109—11, 116—22,
   146
Stationarity   68 (defined), 89
Statistical:
   methodology   18
   significance   18
Steiner, G.A.   127f, 149, 152
Strip-programming   6 (defined), 64, 86f,
   90, 95

156

Swanson, C.E.   49, 152

Talk programmes   5
Television Audience Measurement Ltd.
    (TAM)   vii, xi, 14, 67 (reach and
    frequency guide), 136, 138, 142
TAM Comparison Survey   139
Television, background   1–4
Television sets, number of   1, 3f
Thomas, S.G.   67, 151
Time-bands   19
Times, The (London)   10
Transmissions, start of   1 (UK), 3 (US)
Trivialisation   132
TVQ system   142
Twyman, W.A.   139f, 151

Ultra-high frequency (UHF)   3
United States   2–4, 15f, Chapter 7, 138
US Department of Health, Education
    and Welfare   131, 152

Variety programmes   5
Very high frequency (VHF)   3
Viewer:
    contributions   3
    diaries   7, 137–9
    "gratification"   143
    panels   vii, 7, 136–9, 142f

presence   140–2
satisfaction *see* Audience appreciation
typology   142
Viewers:
    habitual   ix, 13, 15, 130
    repeat, new and lapsed   103–5
Viewing:
    behaviour, US   xi, 15f, Chapter 7
    compulsive   130
    habits   105
    intensity   xi, Chapter 6:
        homogeneous   79 (defined), 82
        other channels   62f
        patterns   x, 10
        hours   ix, 9f, 75–7
Vince, P.   131, 149, 151
Violence, and TV   ix, 131f

WABC   88, 91
WCBS   87f, 91
Week-end viewing   37, 42f, 58f, 85 (US),
    145
Westerns   4, 46f, 129, 132
Williams, R.   4f, 152
WNBC   88, 91
WNET   85
WNEW   85, 87f
WOR   85, 87
WPIX   85, 87

The Authors

G.J. Goodhardt has spent twenty years in business, mostly in marketing research and advertising. In January 1975 he took up his first academic position as Reader in Marketing at Thames Polytechnic.

A.S.C. Ehrenberg has spent fifteen years in industry and has also held academic positions at the Universities of Cambridge, Columbia, Durham, London, Pittsburgh and Warwick. He has been Professor of Marketing at the London Business School since 1970 and is also author of the new statistical text *Data Reduction.*

M.A. Collins has worked in survey research since 1962. He is currently Research Director, Social and Community Planning Research.

All three authors have published widely and are consultants to Aske Research Ltd.